PRINCIPLES AND PRACTICE IN BIOBANK GOVERNANCE

To my mother Margaret, who has taught me patience and compassion in difficult times and to never give up even when the end seems near.
Jane Kaye, 15 October 2009

Principles and Practice in Biobank Governance

Edited by

JANE KAYE
University of Oxford, UK

MARK STRANGER
University of Tasmania, Australia

ASHGATE

Published by
Ashgate Publishing Limited
Wey Court East
Union Road
Farnham
Surrey, GU9 7PT
England

Ashgate Publishing Company
Suite 420
101 Cherry Street
Burlington
VT 05401-4405
USA

www.ashgate.com

British Library Cataloguing in Publication Data
Principles and practice in biobank governance.
1. Biobanks--Management. 2. Biobanks--Moral and ethical aspects.
I. Kaye, Jane, 1962- II. Stranger, Mark.
576.5'0752-dc22

Library of Congress Cataloging-in-Publication Data
Principles and practice in biobank governance / [edited] by Jane Kaye and Mark Stranger.
p. cm.
Includes bibliographical references and index.
ISBN 978-0-7546-7825-0 (hardcover)
1. Tissue banks--Management. 2. Biobanks--Management. 3. Tissues--Collection and preservation--Government policy. 4. Biological specimens--Collection and preservation--Government policy. 5. Procurement of organs, tissues, etc.--International cooperation. 6. Bioethics. I. Kaye, Jane, 1962- II. Stranger, Mark.
[DNLM: 1. Biological Specimen Banks--standards. 2. Clinical Governance. 3. Public Policy. QU 23 P957 2009]
RD127.P76 2009
362.17'83--dc22

2009030145

ISBN 9780754678250 (hbk)
ISBN 9780754696193 (ebk)

Mixed Sources
Product group from well-managed forests and other controlled sources
www.fsc.org Cert no. SA-COC-1565
© 1996 Forest Stewardship Council
FSC

Printed and bound in Great Britain by
MPG Books Group, UK

Contents

List of Figures and Table

Figures

Table

Notes on Contributors

Lisa Austin is trained as a lawyer (LLB [Toronto], member of the Bar of Ontario) and a philosopher (PhD, MA [Toronto]), and is currently an associate professor at the University of Toronto Faculty of Law. Prior to joining the faculty, she served as law clerk to Mr Justice Frank Iacobucci of the Supreme Court of Canada. Professor Austin's current research focuses on theories of privacy, including the relationship between privacy and identity, the role of consent, and the nature of private places. Her work places a particular emphasis on understanding the challenges that new technologies pose to our understandings of privacy and the implications of this for the various legal regimes that protect privacy interests.

Karine Bédard graduated with a BSc in Microbiology from Laval University in 2002 and a MA degree in Bioethics in 2005. Her project looked at the ethical issues perceived by lay populations regarding biobanking. She also completed during year 2007-2008 several courses of the master degree in Genetic Counselling at McGill University. Karine has been a research associate with the *HumGen International* team since 2005. Her tasks are diverse and her research topic focuses on population genetic research, biobanks and public participation. She is currently working with Legal and Ethics Affairs of the CARTaGENE biobank and coordinating the ELSI and Privacy Task Force of the Partnership for Tomorrow Project, a Canadian resource dedicated to health research. Karine has also been a member of the Research Ethics Board of Notre Dame hospital (CHUM) since 2003.

Geoff Bradley is a research fellow of Trinity College Dublin and a part time lecturer in the School of Mathematics, TCD. He completed a PhD in Computational Physics in 2000, and an MSc in High Performance Computing in 1998. Geoff is Executive Director (acting) of the Trinity Centre for High Performance Computing, where he has responsibility for the strategic development of HPC in Trinity, and the overall management of the centre research projects, software development, and services. Dr Bradley has over 10 years experience working in the fields of High Performance Computing and Distributed Data Management. He has been involved in a number of large scale European HPC projects and served on the board of HPC-Europa – a pan European Research infrastructure on HPC. His research interests include algorithm design and development particularly in the fields of molecular dynamics and health informatics.

R. Jean Cadigan, PhD, is an anthropologist with the Center for Genomics and Society at the University of North Carolina at Chapel Hill. Her training is in medical and psychological anthropology as well as human development. Her current research focuses on the ethical, legal, and social implications of genetic research, with particular focus on biobanks.

Anne Cambon-Thomsen is an MD, with an MA in Human Biology and a degree in Health Ethics. She is Director of Research in CNRS and presently leads a multidisciplinary team on 'Genomics, Health, Society', involving also human and social sciences in the context of research in epidemiology and public health. This team is part of an Inserm unit – U558 – entitled 'Epidemiology and analyses in Public Health: Risks, chronic diseases and handicaps' at the Faculty of Medicine in Toulouse. She was a member of the CCNE (National Ethics Advisory Committee in France) from 2002 to 2004. She is currently a member of the Advisory Committee for Priority 1 (Genomics and Biotechnology for Health) of the 6th EU R&D Framework and of the Scientific Advisory Board of Genome Quebec. She has been rapporteur of an expert group at EU level that produced a report and recommendations in 2004 on ethical, legal and social aspects of genetic testing.

Antonio Casado da Rocha holds an MA degree in Philosophy from the National University of Ireland and a PhD in Philosophy from the University of the Basque Country. Since 2002 he has been working at that University's San Sebastian campus as a Research Fellow in the field of bioethics, in collaboration with its Philosophy of Biology research group (www.ehu.es/ias-research), on issues dealing with the concept of disease, health care ethics, research ethics, and environmental ethics. He is a member of the Hospital Donostia ethics committee, and his main fields of work are the concept of autonomy in bioethics, and the regulation of biobanks and biomedical research in Spain.

Denis Chartier is a computer scientist specializing in web-development. Keen on programming in the universe of Web from a young age, he obtained the degree of technology in computer sciences from the University of Toulouse III in 2008. He is currently working at the French National Institute of Health and Medical Research, Unit 558, with Anne Cambon-Thomsen's team in Toulouse. He is involved in a European-funded project as webmaster through the creation and development of computing tools including databases related to exchanges of human biological samples. His paradigm of coding leans on the Symfony framework and he is particularly concerned with problems of accessibility by disabled persons and the issue of standardization.

Christine Critchley is a social psychologist and senior lecturer in the Department of Psychology at Swinburne University in Melbourne, Australia. From 2002-2008, she was the Director of the Public Perceptions Research Unit within the Australian Centre for Emerging Technologies and Society, and is now the Director of the Social Psychology Research Unit at Swinburne University. Her research interests lie in the areas of attitude formation and change, public opinion and trust. She has been involved in a number of multidisciplinary projects exploring public attitudes towards controversial scientific research and trust in scientific actors. This has included designing and conducting six Swinburne National Science and Technology Monitors (SNSTM), an annual national telephone survey designed to gauge public reaction to science and emerging technologies. Dr Critchley has also completed a number of government research projects and reports in the areas of public understanding of science, trust in public sector governance and work place integrity. Her most recent work examines the influence of commercialization on the publics' acceptance of new technologies, including stem cell research, therapeutic cloning, genetic testing, and biobanks. Dr Critchley teaches social psychology, research methodology and advanced quantitative methods at undergraduate and post graduate levels. She also runs research methodology and statistics courses for public servants and government research centre staff.

Arlene M. Davis, JD, RN, is an attorney and assistant professor in the Department of Social Medicine in the School of Medicine at the University of North Carolina – Chapel Hill. She is also a Director of Clinical Consultation and Education for the UNC Hospitals Ethics Committee and a Fellow in the UNC Parr Center for Ethics. Ms Davis received a bachelor of science in nursing with honors from South Dakota State University and a *juris doctor* from the University of Washington School of Law. In conjunction with her law degree, she completed an externship in health law and ethics at the UNC School of Medicine and UNC Hospitals. Ms Davis has been co-investigator on a series of grants from NHGRI's ELSI Program (ethical, legal, and social implications of the human genome project). Most recently, as a co-investigator in the Center for Genomics and Society, she examines the ELSI issues arising in large-sample gene discovery and disclosure. Her research focuses on the creation, understanding, and dissemination of genetic information through genetic screening and biobanking.

Kris Dierickx is professor of biomedical ethics and a staff member of the Centre for biomedical ethics and law, Faculty of Medicine (K.U. Leuven, Belgium). His research is focused on ethics in genetics, research ethics, regenerative medicine, and biobanks. He is partner in several national and international research consortia. Kris Dierickx is currently also the coordinator of the EC funded FP6 project GeneBanC: Genetic bio and dataBanking: Confidentiality and protection of data: Towards a European harmonization and policy (www.genebanc.eu).

Ismael Etxeberria Agiriano is a lecturer in computer science at the University College of Engineering of Vitoria-Gasteiz, University of the Basque Country. He received his degree in Electronics Engineering and his degree in Computer Science from the University of Mondragon (Spain). He received his PhD in Computer Science from Staffordshire University (Great Britain). His main research interests are in Distributed Systems, Computer Security and Bioethics, among many others. Previously, he worked for three years as a consultant for the Crédit Lyonnais in Paris (France) and continued working on financial applications for another three years before returning to education.

Jane Grimson graduated from the University of Dublin, Trinity College, with a degree in Computer Engineering, following which she obtained a masters and doctorate in Computer Science from the Universities of Toronto and Edinburgh, respectively. She holds a Personal Chair in Health Informatics at Trinity College Dublin where she is also Director of the Centre for Health Informatics. She is currently on partial secondment as Director of Health Information to the Health Information and Quality Authority in Ireland. Her main research interests are in health information systems, electronic health records, biobank information management and ICT for development.

Lukas Gundermann is a lawyer (first and second German State Exam, LLM [Edinburgh]). He worked as a research assistant at Universities in Greifswald and Potsdam before joining the Independent Centre for Privacy Protection Schleswig-Holstein (ULD), a regional data protection authority in Germany, in 1996. He provided his expertise and worked at the management level on a number of EU funded projects aimed at implementing a functioning data protection system in different countries. As a Head of Division at the ULD, he oversees several areas, including data protection in the medical field. In this capacity, he has been involved in projects exploring the legal implications of biobanking at the national and European level, in particular with a view to data protection. He has published numerous papers and given a number of talks on data protection, in particular in the medical field.

Kristien Hens holds a master's degree in Applied Ethics, a master's degree in Germanic languages (English and German) and a teacher training degree. She is currently working at the Center of Biomedical Ethics and Law at the Katholieke Universiteit Leuven (Belgium). Her research focuses on the ethics and governance of biobanks. Ms Hens is working on a PhD on the ethical aspects of the storage and reuse of archived tissue from minors for genetic research, under supervision of Professor Dr Kris Dierickx. She is also involved in a national FWO-funded project to investigate current practices and opinions with regard to DNA sample collections in the Belgian Center of Human Genetics. She also participates in the European Commission funded GeneBanC project (FP6).

Adrienne Hunt is secretary to the UK Biobank Ethics and Governance Council, providing administrative and executive support to the committee. Trained as a biochemist, she spent several years working on the Human Genome Project. During this time she was responsible for project and personnel management, including the co-ordination of the sequencing of human chromosomes 22 and 9.

More recently, she worked in academia undertaking research on the ethical issues raised in the context of public health interventions and conducting an investigation of the law and practice relating to the use of patents by governments with a focus on the provision of health care.

Nadja Kanellopoulou is currently a postdoctoral researcher in Law at the Ethox Centre (Oxford). She is a lawyer by training (PhD and LLM, Edinburgh; LLB, Athens; Admission to Practice [Athens Bar]) and was previously an ESRC Genomics Forum Research Fellow (Edinburgh). She specializes in Medical Jurisprudence and Intellectual Property with particular interest in biotechnology law and regulation of individual and group identity. She has expertise in comparative legal and social aspects of human tissue research and assisted reproductive technologies, such as biobanking, tissue donation, benefit sharing, bio-property, public participation. Her current research is focused on legal and regulatory aspects of privacy and consent in the protection of personal data. She is a member of several international law and ethics research networks and journal editorial boards involving the governance of medical and innovative technologies.

Jane Kaye is a lawyer by training (DPhil Oxon; Admission to Practice [ACT Supreme Court]; LLB [Melb.] BA [ANU]) and is the Wellcome Trust Research Fellow in Law, at the Ethox Centre. She is Director of the University of Oxford's Oxford Bioethics Network (OxBioNet) and within the Ethox Centre, she is responsible for directing the Law, Medicine and Emerging Technologies research programme. Her current research is directed at understanding the role of law in the global regulation of innovative technologies in medicine, with a particular focus on biobanks and standardization. She is also active within P^3G, as well as being on a number of editorial boards and expert review panels involving new technologies in genomics and medicine.

Bartha Maria Knoppers, PhD, holds the Canada Research Chair in Law and Medicine. She is a professor at McGill University, Montréal and was formerly at the Faculté de droit, Université de Montréal and senior researcher at the Centre de recherche en droit public (CRDP). Former Chair of the International Ethics committee of the Human Genome Organization (HUGO) (1996-2004), she was also a member of the International Bioethics Committee of the United Nations, Educational, Scientific and Cultural Organization (UNESCO) which drafted the Universal Declaration on the Human Genome and Human Rights (1993-1997) and Co-Founder of the International Institute of Research in Ethics and Biomedicine (IIREB). From 2000-2006 she served on the Board of Genome Canada, and in 2003, became Chair of the Ethics Working Party of the International Stem Cell Forum as well as founding the international Public Population Project in Genomics (P^3G).

Stephanie Lazor, BA, BCL, LLB, is a recent graduate of the Integrated BCL/LLB programme at McGill University's Faculty of Law. While pursuing her legal studies, she worked as a legal research assistant in the Policymaking Core of the Public Project in Population Genomics' International Working Group 3 – Ethics, Governance and Public Engagement at the Centre de recherche en droit public at the Universite de Montreal. Her research interests include studying the legal and policy issues in ethics, governance and public engagement surrounding population biobanks. Other publications include: Wallace, S., S. Lazor and B.M. Knoppers, 'Consent and Population Genomics: The Creation of Generic Tools', *IRB* (forthcoming); Wallace, S., S. Lazor and B.M Knoppers, 'What's in a Clause? A Comparison of Clauses from Population Biobank and Disease Biobank Consent Materials', in Dabrock, P. and J. Reid (eds), *Trust in Biobanking*. Springer: Heidelberg (forthcoming); and Bédard, K., S. Lazor and B.M. Knoppers. 'Mécanismes de gouvernances pour les resources/biobanques populationelles: Recommendations pour le Québec' (2007).

Graeme Laurie is Professor of Medical Jurisprudence in the School of Law at the University of Edinburgh and Director of AHRC/SCRIPT, a law and technology centre in the University sponsored by the Arts and Humanities Research Council. Graeme Laurie served on the Interim Advisory Group established by the funders of UK Biobank from 2002-2004 to advise on governance of the project and he is currently the Chair of the UK Biobank Ethics and Governance Council (2006-2010).

Trudo Lemmens is Associate Professor at the Faculties of Law and Medicine of the University of Toronto, and a member of the Joint Centre for Bioethics and the Centre for Ethics. He holds law degrees from the K.U. Leuven and McGill University (LLM *bioethics*; DCL). He has been a member of the School of Social Science of the Institute for Advanced Study in Princeton (2003-2004), a visiting fellow of the Royal Flemish Academy of Belgium for Science and the Arts (2006-2007), and a visiting professor at the K.U. Leuven, the University of Otago Faculty of Law, and Osgoode Hall Law School. His research currently focuses on the role and impact of law and regulation on medical research, drug development, and biotechnological innovation. He is currently a member of the Advisory Committee on Health Research of the Pan American Health Organization and of the Ethics Advisory Committee of the Canadian Longitudinal Study on Aging. He teaches courses on Medical Law, Research Ethics, and Privacy, Property and the Human Body.

Aurélie Mahalatchimy holds a master's degree in International and European Law. She is currently working at the French National Institute of Health and Medical research, Unit 558 in Toulouse, with the team of Anne Cambon-Thomsen and Emmanuelle Rial-Sebbag. She is involved in two EU-funded projects related to transplantation and cell therapy and exchanges of human biological samples.

She is also working on a national project on stem cells under the supervision of Virginie Tournay for a joint unit of the French National Center for Scientific Research (CNRS) and the Institute of Political Science (PACTE) of Grenoble. She is working on a PhD on 'the impact of the regulation of advanced therapy medicinal products in the Member States of the European Union' in the Faculty of Law, University of Toulouse, France, under the co-direction of Professors Nathalie de Grove-Valdeyron and Florence Taboulet. She is also teaching European Substantive Law at the Faculty of Law, University of Toulouse.

Dianne Nicol is a professor at the Law Faculty, University of Tasmania. She teaches in the areas of intellectual property law, information technology and the law, biotechnology and the law, media law and equity. The broad theme of Dianne's research in the law discipline is the regulation of biotechnology and human genetics. She is particularly interested in the commercialization of genetic knowledge and patenting of genetic inventions as well as the specific regulatory issues associated with genetics, cloning and stem cell technology and biobanking. Dianne started her academic career as a scientist, receiving a PhD from Dalhousie University in Canada in 1987. In 1996 she obtained a first class honours degree in law from the University of Tasmania and in 1997 she completed an LLM on patenting of genetic technologies in Australia. She was admitted as a barrister and solicitor to the Supreme Court of Tasmania and the High Court in 1998. She commenced work as an academic at the University of Tasmania in 2000.

Margaret Otlowski is Law Professor at the University of Tasmania and Deputy Director of the Centre for Law and Genetics (co-located with University of Melbourne). She has been admitted to practice as Barrister and Solicitor and works part-time in quasi-judicial roles, currently as member of the Tasmanian Anti-Discrimination Tribunal. She has worked as consultant for key national bodies including the former National Bioethics Consultative Committee, the Australian Law Reform Commission (ALRC) (inquiry with the Australian Health Ethics Committee into the protection of human genetic information) and member of the ALRC's health sub-committee for its privacy inquiry) and has recently been appointed to the NHMRC's Australian Health Ethics Committee and their Human Genetics Advisory Committee. She chaired the University of Tasmania's Human Research Ethics Committee, is a member of the Royal Hobart Hospital's Clinical Ethics Committee, and of the Federal Privacy Commissioner's 'Health Leaders Forum'. She has been involved as Chief Investigator on a number of Australian Research Council funded collaborative research projects investigating issues of regulation, privacy, consent and genetic discrimination, including a current biobanking project.

Martin Richards is a Leverhulme research fellow at the Centre for Family Research, University of Cambridge and currently a research fellow at the ESRC Genomics Forum in Edinburgh. His research concerns social and ethical aspects of genetic and reproductive technologies. He is also engaged in writing a social history of these techniques. Professor Richards is also Vice Chair of the UK Biobank Ethics and Governance Council and holds an Emeritus Fellowship from the Leverhulme Trust.

Emmanuelle Rial-Sebbag is a graduate in Health Law (Faculty Bordeaux). She has been working since 2000 at the INSERM Unit 558 in Toulouse (Epidemiology and public health analysis: Risks, chronic diseases and handicaps) in the team *Genomics and public health: Interdisciplinary approach* (Dir: Anne Cambon-Thomsen) as a researcher in law and bioethics. She is an associate lecturer in biolaw and bioethics at the School of Medicine, University of Toulouse (Purpan). She is involved in several research projects at the national, European and international level, on the topics of biobanking, transplantation, innovative therapies (Gene and cell therapy) and biomedical research involving human beings. She is responsible for several teaching and instructional sessions as a member of the 'Genetic and Society Platform' attached to the *Génopôle Midi-pyrénées* especially on the ethical and legal aspects of biomedical research involving humans, on ethical aspects of the relationship between medical doctors and patients and on patients' rights regarding biobanking. She is currently developing a new research field on the Governance of Research in biotechnology and the role of regulation at the national and European level.

Loane Skene, LLB (Hons) (Melb), LLM (Monash), LLD (Melb), is a professor of Law in the Faculty of Law and an adjunct professor in the Faculty of Medicine, Dentistry and Health Sciences at the University of Melbourne. She is Deputy Director of the Centre for Law and Genetics, University of Tasmania, and has served on many federal and state policy advisory committees, especially in relation to genetics and the law. In 2005, she was Deputy Chair of the Lockhart Committee on Human Cloning and Embryo Research and became principal spokesperson for the Committee after the sudden death of the Chair, the late Justice Lockhart AO, in January 2006. She is the author of two books on medical law and numerous chapters in books and articles in Australian and overseas legal, medical and scientific journals. In 2003, she was awarded a Centenary Medal for 'Service to Australian Society through the Exploration of Legal and Ethical Issues of Health Care'.

Ulrich Stockter is a lawyer and has worked as an assistant researcher at the University of Cologne, at the German Ethics Committee (Deutscher Ethikrat) and in the bureau as a member of the German Bundestag. In 2007 he joined the Independent Centre for Privacy Protection (ULD) as judicial assistant responsible for data protection in health care and social security. Since mid-2008 he has worked for the German Federal Ministry of Family Affairs, Senior Citizens, Women and Youth. His Doctoral Thesis was on the concepts of legal prohibition of genetic discrimination (2008). He is the author of comprehensive legal opinions concerning the need to inform patients in the field of predictive medicine, especially early detection of cancer (2008) and concerning data protection compliance of biobanks especially with regard to trans-national tissue transfers within Europe (forthcoming 2009). He is co-author of data-protection regulations in a legal commentary on German Transplantation Law (2009).

Mark Stranger is a senior research fellow with the Centre for Law and Genetics at the University of Tasmania, where he has worked since 2002. He is the Executive Director of the Centre and Manager of the Centre's multidisciplinary, international Biotechnology, Ethics, Law and Society (BELS) Network. Mark has a PhD in sociology and his interests include risk, science and technology, and theories of social change. He is a member of the University of Tasmania's Research Peer Review College and a member of the University's Internal Research Grants Scheme review board. Mark is the Project Manager for the Centre's current ARC Discovery Grant project, 'Facilitation and Regulation of Research and Development Involving Human Genetic Databanks'.

Mark Taylor is a lecturer in Jurisprudence and Human Rights and Human Genetics at the University of Sheffield. Appointed in 2002, he has an LLB from the University of Hull and an MA in Biotechnological Law and Ethics and a PhD from the University of Sheffield. He is a member of SIBLE (Sheffield Institute for Biotechnological Law and Ethics) and the Ethics and Confidentiality Committee of the NIGB (National Information Governance Board for Health and Social Care). He is currently joint Principal Investigator on an EC project called PRIVILEGED (Privacy, Law, Ethics and Genetic Data). PRIVILEGED considers the ethical and legal interests in privacy and data protection for research involving the use of genetic databases and biobanks (http://www.privilegedproject.eu/).

David Townend is head of the law group and Associate Professor of the Law of Public Health and Care within the Department of Health, Ethics and Society in the Faculty of Health, Medicine, and Life Sciences in Maastricht University. Prior to his appointment in Maastricht, he was a senior lecturer in the Department of Law and Sub-Dean (Postgraduate Studies) in the Faculty of Law in the University of Sheffield. He was also a Deputy Director of the Sheffield Institute of Biotechnological Law and Ethics (SIBLE). His major research interest lies in property rights and their relationship to human rights, particularly in relation to intellectual property, to data protection, privacy, medical research, genetic information and biobanking, and to social welfare. He has co-ordinated and participated in many funded research projects, including work funded under the EC FP5, 6 and 7 frameworks. He worked with Professor Deryck Beyleveld (Durham) in the co-ordination of the PRIVIREAL project which examined Europe-wide the implementation of the Data Protection Directive (95/46/EC) in relation to medical research and its operation in research ethics committees. He is co-ordinator (with

Dr Mark Taylor) of PRIVILEGED which examines issues, again Europe-wide, of privacy and data protection in relation to genetic information and biobanking.

Susan Wallace is a visiting researcher at the Centre de recherche en droit public (CRDP), Université de Montréal, Montréal, Canada. She is Director of the CRDP Policy-making Core which examines policy issues related to population biobanks. The Core works with the International Working Group on Ethics, Governance and Public Engagement of the Public Population Project in Genomics (P^3G) Consortium to create tools for the biobanking and academic community. Her research interests include the policy implications of population-based and disease-based biobanking, the ethics review of human subject research, public health ethics and public health genomics. Immediately prior to joining the CRDP, Susan was Policy Officer (Humanities) at the Public Health Genetics Unit, Cambridge Genetics Knowledge Park, in Cambridge, UK. She has also been the Director of the Americas Office of the Human Genome Organisation in Bethesda, Maryland, USA. She obtained her PhD in 2004 at the University of Sheffield, Sheffield, UK, in Biotechnological Law and Ethics. She is also currently Associate Editor of the journal, *Public Health Genomics*.

William Watson is Associate Professor of Cancer Biology in the UCD School of Medicine and Medical Science, University College Dublin. Professor Watson received his PhD degree in Biochemistry from the Department of Biochemistry, University College Cork in 1995 after undertaking his PhD studies in the Department of Surgery, Royal College of Surgeons in Ireland. He then undertook his post-doctoral research in University of Toronto and the Toronto General Hospital in Canada, before returning in 1997, to Ireland as a College Lecturer in the Department of Surgery, Mater Misericordiae University Hospital, University College Dublin and then Senior Lecturer in the UCD School of Medicine and Medical Science from 2004-2007. Apart from his undergraduate and post-graduate teaching responsibilities, he has an active research group in the field of prostate cancer biomarker discovery and therapeutic manipulation. As a translational biologist based in the UCD Conway Institute of Biomolecular and Biomedical Research, he utilizes latest technologies in cellular and molecular biology and clinical collaboration through the Prostate Cancer Research Consortium, of which he is Co-Principal Investigator and Chair of the Bioresource management and implementation committee, to expand the understanding of the initiation and progression of prostate cancer.

David Weisbrot, AM, Emeritus Professor at the University of Sydney, is President of the Australian Law Reform Commission. He has chaired major national inquiries into the protection of human genetic information (the Essentially Yours report, 2003), gene patenting and human health (Genes and Ingenuity, 2004), and privacy laws and practice (For Your Information, 2008), and has published extensively in areas relating to the ethical, legal and social implications of genetic research, clinical practice and biobanking. Professor Weisbrot is a member of the Human Genetics Advisory Committee of the National Health and Medical Research Council, and a Fellow of the Australian Academy of Law. In 2006, he was made a Member of the Order of Australia for services in the areas of law reform, education, access to legal services and policy development on matters of public interest.

Dita Wickins-Drazilova was a research assistant (at the time of writing her chapter), at Sheffield Institute of Biotechnological Law and Ethics (SIBLE), University of Sheffield, working on the Privacy, Law, Ethics and Genetic Data (PRIVILEGED) project (http://www.privilegedproject.eu/), looking at the ethical and legal interests in privacy and data protection for research involving the use of genetic databases and biobanks. Dita now works at the Department of Philosophy, Lancaster University, on the IDEFICS project (http://www.lancs.ac.uk/fass/philosophy/activities/226/), which is concerned with the ethical and public policy implications of interventions into childhood obesity. Dita's main research interests are in applied ethics, specifically bioethics, ethics of childhood, public-health ethics, research ethics, animal ethics and environmental ethics. Her PhD is from Masaryk University, Brno, Czech Republic and part of the research was conducted at Colorado State University thanks to a Fulbright scholarship. She has an MA in Environmental Studies and Sociology from Masaryk University, and an MA in Philosophy from Charles University, Prague.

David E. Winickoff joined the U.C. Berkeley faculty in 2004 after teaching at the John F. Kennedy School of Public Policy at Harvard University, and after degrees from Yale University, University of Cambridge (Paul Mellon Fellow), and Harvard Law School. Drawing upon law and STS, his bioethics work spans topics of genomic governance, stem cell research, property and intellectual property, race, environmental regulation, international trade, and university-industry relations. He has published 29 articles in leading bioethics, biomedical, law and science studies journals such as *The New England Journal of Medicine*, the *Yale Journal of International Law*, and the *Journal of Law, Medicine, and Ethics*. Winickoff was a visiting professorial fellow at the University of Edinburgh in 2006, and a visiting scholar at the Kennedy Institute of Ethics at Georgetown in 2008. He is currently the Co-Director of the Science, Technology and Society Center at U.C. Berkeley, where he has organized and chaired numerous conferences on the bioethics of intellectual property in the life sciences, drawing large national audiences. In 2007, Winickoff was chosen to be a Greenwall Faculty Scholar in Bioethics, a national award in the United States, for work on distributive justice and University licensing in the life sciences.

Jessica Wright works as a research co-ordinator on two European Framework 6 projects, 'PRIVILEGED' (University of Sheffield) and '@neurIST' (Durham University). PRIVILEGED examines issues of privacy and data protection in relation to genetic information and biobanking and @neurIST aims to integrate biomedical informatics tools to better manage cases of cerebral aneurysm. The more theoretical work on PRIVILEGED is matched by the practical application of ethico-legal considerations on an actual medical research project, which includes the collection of DNA samples and the creation of a Biobank. She also studies part-time for a PhD in the Department of Law, University of Sheffield.

Atieh Zarabzadeh obtained her bachelor degree in Software Engineering from Azahra University, Tehran, followed by a postgraduate diploma in Health Informatics from the University of Dublin, Trinity College. She is currently working on a PhD in Computer Science at Trinity College Dublin. Her thesis is concerned with sample identification tracking for multi-location or networked biobanks and in particular on the use of Radio Frequency Identification (RFID) technology.

Acknowledgements

This book is based on key papers presented at an international conference held in Oxford, UK, 24-26 June 2008 on the topic of governing biobanks. We believe that this was the first conference of its kind to focus specifically on this topic. The conference brought scholars from 34 countries from across the world, in order to present and develop new insights in this field and to stimulate further research. It was co-convened by Jane Kaye of the Ethox Centre at the University of Oxford and Mark Stranger of the Centre for Law and Genetics University of Tasmania, Australia. The idea for the conference grew out of the *Governing Genetic Databases Project* at the Ethox Centre, funded by the Wellcome Trust (2005-2009; WT 076070/Z/04/Z), and the Australian Research Council funded project, *Facilitation and Regulation of Research and Development Involving Human Genetic Databanks* (2005-2009; DP0559760), based at the Centre for Law and Genetics.

The conference was generously funded by the Wellcome Trust (WT 085239/Z/08/Z and WT 076070/Z/04/Z) and the Australian Research Council (DP0559760). We were able to provide travel assistance to people from Africa, South America, India, Australia, Canada, Singapore, Taiwan, Japan, USA, Norway and Russia. We were fortunate that student bursaries for the conference were provided by the European Science Foundation (Ref 1737).

This book would not have been possible without the editing skills of Helen James (Centre for Law and Genetics) and the administrative skills of Imogen Holbrook (Ethox Centre). We would like to thank them for their hard work, perseverance and for being so good-natured at all times! We would also like to thank the authors of the chapters in this volume, for the presentation of their papers at the conference, but also for the huge amount of effort that they have taken to turn their presentations into enduring words on the page.

Jane Kaye and Mark Stranger

Governing Biobanks: An Introduction

Mark Stranger and Jane Kaye

The collection and storage of human tissue for medical research is not a new phenomenon. Neither is medical research interest in genetic heritage new, although before the relatively recent advances in genetic science and technology this interest was expressed in terms of family history. A medical practitioner could expect questions about a patients' family medical history to be taken as a sign of their thoroughness and care. Such inquiries are perhaps accepted as benign or benevolent because they are like a natural extension of common civil greetings; such as 'How are you?' and 'How is your family?' But with the rapid advances in the human genetic sciences and the emergence of biobanks (and their potential links to other databases) the medical practitioner's inquiry could end up sounding more sinister.

No don't talk. Just open your mouth wide while I take a swab.
I'll be able to find out all I need to know about your family history ...
and its future.

A genetic test has none of the benign veil of civility that cloaked the intent of a trusted doctor's inquiry into our family's medical history, rather its shadow can loom large with the fear of discovering some dreadful and inevitable future, and the potential for discrimination.[1] The human genome's 'Book of Life' can be seen as just the introductory volume in an infinite series, where every individual has the 'Book of *MY* Life' written in their genes and awaiting publication. While the fear of uncovering indications of future diseases or disorders and concerns over the implications of acquiring this knowledge may lurk in the minds of the public, there is still a market for personal genetic information, as the emergence of direct to consumer testing companies demonstrate.

Rather than going through your local doctor, it is now possible to carry out a buccal swab test in the comfort of our own homes and log onto an internet account with a direct to consumer testing company. We can even have the results downloaded onto our iphones. The activities of direct to consumer testing companies are changing the landscape of genomics rapidly. Sequencing is no longer the exclusive domain of specialist scientists. These companies are here to stay, even if the information on risk that they offer is currently of limited clinical

1 http://www.gdproject.org.

utility. It is therefore more important than ever that the research community clearly articulate the principles that distinguish and underpin good practice in the field.

Current scientific opinion is divided over whether genetic information has any significance for many common diseases, as it is difficult to replicate findings and a lot more work needs to be done to determine the complex and subtle relationship between genotypes, phenotypes and environment (McCarthy et al. 2008, Ioannidis et al. 2009). The need for large sample sizes to understand and measure these complexities (Burton et al. 2009) has fuelled the drive to establish new biobanks, but also to link existing collections of samples and information. The perceived importance of biobanks is evidenced by the funding initiatives to develop infrastructure that can link and integrate biobanks and research collections. However, support of the public is essential, as it is only with continued donation of samples and data, and large-scale investment programmes that genomic research will continue. Good governance of biobanks is where we can first and most comprehensively address some of the fears and concerns of the public. It is well accepted that good governance structures for research have the potential to encourage greater public trust.

There are many different kinds of biobanks, both clinical and research focused, and these facilities can be called by different names (e.g. human genetic databanks, human tissue collections, biospecimen repositories, DNA Databases) (Tutton and Corrigan 2004, Cambon-Thomsen et al. 2007, Kaye 2006). In its most basic form, a biobank can be described as a collection of human tissue and the genetic information collected from that tissue. Combined with other sources of personal information biobanks can be used for different kinds of research by many different researchers. The value of the biobank is largely contingent upon being able to link the samples with donor information and its value increases with the depth and quality of the information. Biobanks can facilitate continuous collection of data and information over extended periods of time which maximizes the value of the existing resources and the potential to discover more about medical conditions. In an ideal world – for the researchers at least – the information linkages would include ongoing access to the donors' medical records, and information about the environmental and behavioural risks that characterize their everyday lives. The list of potential contributing environmental and behavioural factors and combinations is endless. And so, it would seem, is the endeavour.

The term *governance* when applied to biobanks can have different meanings. When considered in terms of 'corporate' governance it relates to the business structures set up to manage the institution and ensure it achieves the desired outcomes within the relevant systems of regulation. Such governance frameworks have been described as, '… the agreements, procedures, conventions or policies that define who gets power, how decisions are taken and how accountability is rendered' (Institute on Governance 2003 in Bédard et al. this volume). Therefore a governance structure is a combination of procedures declared through guidelines and policies, and enacted through decision-making bodies. The way that a governance structure is organized and where the control for decision-making resides can fundamentally change the nature of a biobank, its priorities and the way the biobank interacts with the society of which it is a part.

It would be reasonable to assume that the primary focus for those wishing to establish biobanks is to create a system of governance that delivers in terms of outcomes: i.e. the quality of the collection and storage, its desirability as a research resource, the integrity of the system that manages access to specimens and data, and the efficiency and effectiveness of accountability, auditing and security mechanisms. This 'outcomes' focus is apposite to the model of corporate governance. Bioinformatics is considered key to achieving good outcomes-focused governance. However, the chapters in this collection demonstrate that a key concern of many governance structures established for biobanks is to ensure that research is carried out ethically, and research participants are not harmed by involvement in the biobank. But many of the governance systems in place do not involve research participants in decision-making.

As Gottweis and Petersen (2008: 8) point out, the use of the term *governance* represents a blurring of boundaries such that the regulation of biobanks is now an issue within the public arena. Biobank outcomes are contingent upon input in terms of tissue and data and donors must be forthcoming with deposits if the biobanks are going to be able to do their business. The link between donor participation, public trust and public consultation has been widely recognized and public engagement has been embraced as a necessary step in establishing a biobank (Weisbrot in this issue; Levitt and Weldon 2005, Stranger et al. 2005). Public engagement is considered the key to achieving good outcomes in this broader model of governance, and evidence of good corporate governance feeds into the communicative process. It is this broader sense of governance that engages ELSI scholars and is therefore the focus of this publication.

The enquiry into the ethical, legal and social implications (ELSI) of human genetic research and development has been a showcase for multidisciplinary collaboration. But it is a showcase that can never provide the kind of definitive outcomes that those wishing to establish biobanks might hope for. Despite the *need* for harmonization of governance structures in the global research environment (Bovenberg 2005), the 'Book of Life' is unlikely to be matched by the 'Book of Biobank Governance', at least not in the genre of a one-size-fits-all instruction manual. This, in the end, will be a political publication with multiple and varied versions.

The authors that have contributed to this volume range across disciplines and speak from scholarship and experience embedded in different cultures and politico-legal systems. These combined works present what we believe to be a snapshot of key principles and current practices relating to the governance of biobanks across the globe at this point in time. The governance structure of any biobank will be the result of political deliberations that reflect the cultural and political reality in which they are embedded. Any generic blue-print can only be in abstract terms. The best we can hope to offer is basic principles that can be adapted to suit the context. As will become clear, there is still plenty of room for debate around these principles and how best they should be translated into governance structures, policies and guidelines. But this does not represent a failure on the part of ELSI scholarship. Ongoing debates around widely accepted principles and how they should be put into practice are surely indicative of some kind of end-point rather than inertia or confusion; they provide material for public political deliberation that must inform and, to a significant extent, *constitute* the governance of biobanks.

Themes and Chapters

The chapters in this book have been divided into four key themes; Benefit Sharing, Consent, Privacy and Access to Data, and Governing Bodies. What follows is a brief outline of the chapters as they appear, ordered under the themes that are their *primary* focus. As might be expected, given the contingent nature of the issues involved, individual chapters can also range across themes. The reader should therefore not consider the thematic ordering of the chapters as rigidly delineating all discussions of relevance to these topics.

Benefit sharing

Perhaps the notion of a common human heritage that became attached to the human genome has had some influence, but whatever the reasons, benefit sharing has emerged as a key theme in discussions about the governance of biobanks. The concept covers a fairly broad field, ranging from the individual to the global sphere; including ideas about benefits to donors, benefits to third world counties, and arrangements for sharing research data with other researchers. The principle involved here is that there should be some equity in the distribution of goods that accrue from the establishment of biobanks and the genetic research that they facilitate. It appears that we are not content to leave it to the invisible hand of the market to determine how the benefits are to be shared. How then might we best build mechanisms into the governance of biobanks that will help lead to the equitable distribution of benefits? The chapters in this section make a valuable contribution to our deliberations on this issue.

The findings of Nicol and Critchley's research (Chapter 1) show that Australians have clear expectations that there should be benefits to the broader community from their participation in a biobank. There is also a high level of support for: keeping research open, both by returning results to the biobank and by public dissemination; and ensuring that healthcare products developed using biobank resources are affordable to both the Australian and global community. The key conclusions that the authors make in this chapter are that people care about benefit sharing and want to know what benefit sharing arrangements are in place before deciding to participate in biobank research and therefore that as a matter of course, benefit sharing should be explicitly addressed in biobank governance frameworks. The warning for biobanks (especially for those with industry links) is that they risk losing public support if benefit sharing is ignored. Further, the authors argue that public consultation is necessary 'as it assists in understanding the motivations of members of the public to participate in biobanking and how regulatory and governance frameworks must be shaped to satisfy their needs and allay their concerns'.

In Chapter 2, Kanellopoulou provides the conceptual basis for empowerment that could underpin the empirical findings of Nicol and Critchley on benefit sharing. Kanellopoulou argues that the relationship between biobanks and their participants should not be founded solely on the basis of altruism and a gift model but that participants' interests would be better served by the law if an approach was adopted that focused 'on the nature of their interaction with researchers and describe it as an ongoing cooperation and dynamic relationship with special obligations for both sides'. She considers that this shift is necessary because of the increasing economic value of human tissue and commercialization, by both public and private institutions, and because of the long-term participation and commitment undertaken by participants. This requires a conceptual shift to recognize 'the continuous nature of their relationship as dynamic and worthy of protection'. Under this approach, research gifts should be considered as conditional and regulated in accordance with requirements of reciprocity.

Winickoff, in Chapter 3, advocates the notion of 'partnership governance' for biobanks. He argues that biobanking projects are likely to fail both as a normative and practical matter without greater attention to issues of procedural justice: in particular, the constitution of distributive agency over resources for genomic research. For Winickoff partnership governance 'presents a productive avenue for achieving a normative shift from "benefit sharing", a distributive value, to "power sharing", a procedural one'. He asks why the 'principles and legal forms derived from charitable trust law and corporate governance, such as trusteeship, fiduciary duties and shareholding', should not be brought to bear in the realm of charitable biobanking. His idea of 'partnership governance' does not mean exclusive ownership by volunteer individuals, nor does it allow self-interested action. Rather, partnership governance is based on an idea of cooperative human relations forming and acting out different parts of a complex governance structure for public benefit. 'If implemented, partnership governance would empower participants to exert a share in distributive decision-making in return for contributing to the economic and social capital of the project' and 'go further than existing mechanisms of "community consultation", by implementing a share of control at the level of the research participant collective'. Winickoff draws on legal principles and norms from charitable trust law and corporate law to help arrive at partnership governance.

Consent

A requirement for informed consent is regarded as a fundamental principle in medical research ethics. However, biobanks raise considerable difficulties for the practical application of this principle. In particular, problems arise because the samples being donated can remain in storage as a valuable resource for future research unimagined at the time of donation. In practice, the idea of being adequately informed has always been problematic, and in the context of biobanking it is clearly even more so. Given the potential benefits that could be gained from biobanks and the difficulties involved in following the conventional model of informed consent, different approaches have been considered, ranging from forms of re-consent through broad consent and alternative governance mechanisms, and combinations thereof.

In Chapter 4, Gundermann and Stockter argue that at a minimum, broad consent has to be obtained from the donor before their tissue and related data can be processed, but this must wherever possible be accompanied by the possibility that donors can exercise 'co-determination' as a means of safeguarding their autonomy. 'Co-determination' has its origins in German labour law and denotes 'the transfer of democratic decision making processes into private or public management structures'. In the case of biobanking, Gundermann and Stockter regard that 'co-determination' would entail the possibility of donors exerting influence on the precise use of their data and tissue in a biobank and particularly in a population biobank. They argue that the original consent could still be broad and open regarding research uses, but co-determination does however require transparency for donors: 'they must be kept updated on current and envisaged future use of the material'. In this conceptual framework the donor is not considered as 'an indifferent supplier of research tissues but as a partner, who by his or her donation enables research'. Gundermann and Stockter see that the principle of co-determination drawn from law would require the biobank to keep the donors informed about these new developments. This approach is very similar to Winickoff who uses corporate law models of shareholders to argue for greater rights for biobank participants.

In Chapter 5, Otlowski argues in favour of a broad consent for biobanks, by proposing a hybrid model that involves 'informed and specific consent for the collection and storage of samples and information, and broad consent for future use for presently unspecified research'. She argues that consent needs to be reconceptualized in the context of biobanks. She says that although all future research projects cannot be specified at the time samples and information are collected, the core principles consent aims to uphold can and must be protected. When being asked for a broad consent individuals must have 'enough information to understand the general nature of what the research is about'. Otlowski believes that broad consent operates best in 'a well regulated and protective environment augmented by a strong governance regime: indeed, it would be a mistake to rely on consent as the sole basis of protecting research participants'.

In the new Spanish law on biobanks, as described in Chapter 6 by Casado da Rocha and Etxeberria Agiriano, there has been a move towards a 'flexible, middle way' in between a broad consent and informed consent. This Act allows for the possibility that individuals might give explicit consent to the use of their samples for one kind of research project and then consent to further unspecified uses of the samples in projects that are related to the original research project, whether by the same team or another. It is up to a Research Ethics Committee supervising the biobank to make the decision on the unspecified research on the donor's behalf. Although this 'governance-by-committee' approach is still being developed, there are already many exceptions to consent for biobanks in the new Spanish law, and Casado da Rocha and Etxeberria Agiriano have concerns that it might prove to be a slippery slope whereby the requirement for consent is progressively eroded away.

Hens and Dierickx question whether children should be excluded from genetic research and the storage of their samples in pediatric biobanks. In Chapter 7 they acknowledge that the storage of tissue from minors and subsequent research conducted using this tissue, raise ethical issues that do not arise for adults. In particular: (1) the question of consent; (2) the strict requirement in pediatric research that research should pose no more than minimal harm; and (3) that there should be direct benefits to either the child or the group to which he or she belongs. Hens and Dierickx argue that with proper policies and procedures that take these issues into account, there is no need for a ban on involving children in biobanks or genetic research.

In Chapter 8, Cadigan and Davis present findings from a study of healthy volunteers who were approached to enrol in the Environmental Polymorphisms Registry (EPR), based in North Carolina in the US. What is particularly interesting about this study is that it involved interviews both with people who had agreed to enrol in the study, as well as those who had declined to be involved. The chapter discusses the importance of financial incentives and existing governance mechanisms to those who chose to donate a sample of blood to the biobank. However, for those who chose not to participate in the EPR, financial rewards and knowledge about governance safeguards, such as research ethics committee approval, Certificates of Confidentiality, and institutional reputation, did little to allay their concerns about privacy and involvement in a longitudinal genetic research study.

Privacy and access to data

Ensuring the privacy of donors, their relatives and genetic communities, remains a central issue for the governance of biobanks. Regardless of where we might stand on the uniqueness of genetic information and the privacy rights of individuals versus more utilitarian principles, the potential for donors to be harmed through the disclosure of sensitive genetic information is widely accepted. The principle driving developments in this area then is not simply the protection of individual privacy *per se*; it is fortified by the principle that donors and their genetically significant others should be protected from harm *caused* by breaches of privacy. The problem of protecting these people from harm is compounded by the need for samples to be identifiable and linked to other personal information. Providing researchers with access to samples and data in a form that is both useful to them and protects the privacy of donors is problematic; it is further complicated when the researchers work under different governance structures and within different regulatory frameworks. The chapters in this section explore a range of problems and potential solutions associated with the protection of privacy in biobanking.

Townend, Taylor, Wright and Wickins-Drazilova report on some of the preliminary findings from the PRIVILEGED Project. The project is looking at the ethical and legal interests in privacy and data protection in relation to biobanks and genetic databases. In Chapter 9 they discuss the results of a review of research literature from across Europe looking at the public's attitudes and opinions in relation to genetic databases and biobanking. Townend et al. conclude that while there are typically majority positions, these are expressed within a wide range of attitudes and opinions. Simply allowing the majority views to dictate policy for biobank governance, they suggest, would result in the alienation of significant minorities, and that this could undermine the potential value of biobanks. While they recognize the impossibility of encompassing all opinions within a regulatory framework, they maintain that 'an iterative process can be imagined that would, if undertaken, progressively identify the widest range of options for participants consistent with effective research'.

In Chapter 10, Skene considers 'the nature, extent and enforceability of the legal duties' that custodians of biobanks, researchers and other biobank users may owe to donors and perhaps even their relatives, in the case of 'significant findings'. She regards significant findings as those where: '(1) the risk for disease is significant; (2) the disease has important health implications (i.e. fatal or substantial morbidity); and (3) there is a proven therapeutic or preventative intervention available'. Skene discusses Australian law and whether an obligation of feedback exists under statute, contract law, the duty of negligence, the doctrine of confidentiality, privacy statute or property rights. She concludes that it is unlikely that a legal obligation would arise, though she believes that a biobank would have a duty to consider how to make information available under the National Health and Medical Research Council guidelines, rather than to directly warn research participants at risk.

The chapter by Zarabzadeh, Watson, Bradley, and Grimson (Chapter 11) describes the measures that have been put in place for the Irish Prostate Cancer Research Consortium (PCRC) biobank. The PCRC research biobank stores tissue, blood urine and DNA that has been used for diagnosis and determining clinical care. The biobank's infrastructure enables research to be carried out using the samples and information held within the consortium. This is an example of where the divide between the clinic and research has been bridged for the benefit of cancer patients and the wider community. Zarabzadeh et al. maintain that infrastructure design and the standard operating procedures have been developed in accordance with data protection principles, in an effort to ensure the highest standards of confidentiality. The PCRC biobank is said to provide a model that could be used for biobanks elsewhere in order to enable researchers to make greater use of such resources.

In Chapter 12, Rial-Sebbag, Mahalatchimy, Chartier, and Cambon-Thomsen report on an internet-based research tool called hSERN. This tool is designed to provide researchers with the information required to transfer human biological materials (HBM) between France and the United Kingdom (to be extend to other European countries). 'While biobanks are designed to make HBM more accessible to a large number of researchers, they do not necessarily address the procedural issues of how to transfer HBM between biobanks located in different countries'. hSERN came out of the recognition that regulations affecting the movement of HBM across national borders can be a major stumbling block to research progress. hSERN enables researchers to collaborate and can underpin biobanking activities. The different legal requirements that apply within Europe are a serious obstacle to the network of biobanks described by Kaye (following chapter). This tool has the potential to bring significant benefits to researchers, the development of infrastructure as described by Zarabzadeh et al. and the networking of biobanks in general.

In Chapter 13, Kaye starts to explore the issues raised by the development of European infrastructure that will lead to large-scale linkage and sharing of samples and information that are held in biobanks. Such infrastructure, that are still in the process of being developed, would enable researchers to be able to access a number of biobanks at the same time through one portal rather than approaching each biobank for access individually. Kaye sees this as the next logical step from the development of common standards and procedures by organizations such as the Public Population Project in Genomics (P^3G). However, such initiatives create a number of challenges, particularly for the protection of privacy and informed consent. The difficulty with such proposals is that the legal framework and the governance bodies are nationally based and are currently unable to govern infrastructure of this sort.

Governing bodies

While we strive towards some level of standardization, the governance structures of biobanks are necessarily context specific. As each new biobank is established those involved in its development can draw on the experiences of those before them, adopting and adapting principles and practice that suit the particular aims of their facility and the legal, social, cultural and political environments in which they will operate. As noted above, this environment will increasingly include networking between biobanks and even across national borders. But the governance of biobanks is not restricted to the in-house mechanisms of the facilities. It also reaches out into the community in various ways – through ethics committees, public engagement mechanisms and regulatory bodies. A key role of governing bodies is to interpret accepted principles through policies and guidelines, and ensure they are put into practice. The chapters in this section provide an overview and critique of some of the models of biobank governance emerging around the globe.

The chapter by Bédard, Wallace, Lazor and Knoppers (Chapter 14) discusses a study of governance systems for population biobanks that were a part of the P^{3}G consortium. The research was conducted in order to provide a basis for the design of the governance structures for the CARTaGENE population biobank in Quebec, Canada. This study shows that these biobanks are embedded in a complex regulatory framework involving six 'elements' or sets of governance procedures: (1) scientific evaluation; (2) ethics evaluation; (3) data protection and public health laws, and laws on statistics; (4) bio-security standards for laboratories; (5) guidelines for research with human participants; and (6) professional guidelines. Despite these common elements, the study found that there were also significant differences between the governance structures that reflected the particular context in which the biobanks operated. Bédard et al. concluded that '[n]o one governance framework could be determined that would fit all biobanks'. However, they suggest that in the future it may be possible to arrive at an 'optimal governance framework for a population biobank'.

In Chapter 15, Richards, Hunt and Laurie provide a description of the UK Biobank's Ethics and Governance Council (which is one of the committees included in the study conducted by Bédard et al.). This Council was established as an addition to the existing UK research governance and regulatory systems because of the nature of UK Biobank as a very large scale and complex project with very broad purposes and long-term aims. Unlike research ethics committees it can undertake continuing monitoring. And the Ethics and Governance Framework with the independent Ethics and Governance Council as its guardian provide a further safeguard and a foundation for trust. According to Richards et al. 'a body such as the EGC can add value to a governance model by actively monitoring and developing alongside the scientific project'. In the case of the EGC, it has taken on a number of different tasks, such as commissioning research, tracking the concerns of research participants, and providing ethical advice to the UK Biobank.

In Chapter 16, Lemmens and Austin discuss how a combination of various developments makes it increasingly difficult to protect privacy through informed consent procedures alone. They discuss how the concept of Fair Information Practices (FIP) offers an interesting model for a more coherent governance system of health information. This model has already been introduced in the context of research by various Canadian provinces. They say that the promotion of the integration of FIP into the existing system of research ethics review has been done 'without appropriate recognition of the structural weaknesses of the current regulatory regime surrounding research'. This development is particularly important in the context of biobank governance as health information privacy legislation will govern many of the aspects of the collection, use and disclosure of information associated with these biobanks. These privacy statutes do not provide a detailed regulatory framework for the research ethics boards that have to promote and enforce the Fair Information Practices. Lemmens and Austin argue that research ethics boards, as currently regulated, are inappropriate to carry out this activity because the existing ethics review structure does not fulfil established minimum criteria for good governance. They go on to describe an alternative governance model that has been developed in British Columbia (Canada), involving Data Stewardship Committees. These committees have a number of lay representatives, a legislative base and are clearly accountable to the government.

Weisbrot's Chapter 17 demonstrates the important role of government bodies in carrying out community consultation exercises and how these exercises provide a basis for the development of recommendations to government. This chapter provides an insight into two successful public consultation processes held in 2003 and 2008, conducted by the Australian Law Reform Commission that dealt directly with the ethical, legal and social implications of human genetic research and the governance of biobanks. The consultation processes identified a range of concerns that fell into four broad categories: ethical oversight; biobank governance; commercialization, access and equity, and benefit sharing; and genetic discrimination. He argues that 'policy-makers must address the associated, and very legitimate, concerns expressed by members of the community, or else risk a backlash that would dramatically set back all such research, however well-designed and intentioned'. Weisbrot concludes that there is a clear need for openness and transparency if biobanking and population genetics initiatives are to benefit from public confidence. He says experience to date clearly shows that taking ethics and governance seriously is essential for achieving public acceptance and legitimacy, and inherent in this is engaging in community consultation early and on an ongoing basis.

Conclusion

As biobanks flourish across the globe in response to the demands of research and commerce, they do so largely ahead of adequate regulatory frameworks, transcending national borders, pushing the boundaries between public and private enterprise models, pioneering new forms of governance, and embracing the *need* for public engagement (if not the democratic principles behind it). It is essential that those of us engaged in the ELSI of biobanks keep abreast of these global developments. It is to this end that this volume and others like it serve as an important resource.

The themes that have given a framework for the discussions presented here represent key areas around which basic principles for biobank governance have been forming. While debates continue to define and refine the principles in abstract, biobanks themselves are shaping them through practice.

Advances in technology and the momentum towards the integration of collections, and the sharing of data and tissue between biobanks, research teams and across legal jurisdictions, all contribute to an evolving research environment. Governance structures risk becoming redundant if they fail to incorporate mechanisms that facilitate timely adaptation to changing circumstances. Well designed and implemented public engagement processes (as opposed to public relations programs) that are integrated into governance structures, have the potential to function as mechanisms for adaptation.

Public engagement should not be considered as simply a necessary step on the way to establishing a biobank – a box to be ticked and move on – rather it is essential that it becomes an integral part of the overall governance structure of the biobank. This is especially the case given the rapidly evolving nature of the science and research environment and the fluid societies in which they operate.

A biobank governed by regulations, policies and guidelines (informed by public consultation), that is engaged in open ongoing dialogue with its publics, stands the best chance of fostering and maintaining the public trust and support necessary for its successful operation. But even given this ideal, the governing bodies involved will need to steer a path through sometimes competing principles – the principles of good science and those of a liberal democracy are not always in harmony. The chapters that follow do not provide definitive answers to the problems of biobank governance, but they do contribute substantially to the ongoing dialogue that must accompany the organic process that is biobank governance.

Bibliography

Bovenberg, J. 2005. Towards an international system of ethics and governance of biobanks: A 'special status' for genetic data? *Critical Public Health*, 15(4), 369-384.

Burton, P.R., Hansell, A.L., Fortier, I. et al. 2009. Size matters: Just how big is BIG? Quantifying realistic sample size requirements for human genome epidemiology. *International Journal of Epidemiology*, 38(1), 263-273.

Cambon-Thomsen, A., Rial-Sebbag, E. and Knoppers, B.M. 2007. Trends in ethical and legal frameworks for the use of human biobanks. *European Respiratory Journal*, 30, 373-382.

Gottweis, H. and Petersen, A. 2008. *Biobanks: Governance in Comparative Perspective*. London: Routledge.

Ioannidis, J.P.A., Thomas, G. and Daly, M.J. 2009. Validating, augmenting and refining genome-wide association signals. *Nature Reviews Genetics*, 10, 318-329.

Kaye, J. 2006. Do we need a uniform regulatory system for biobanks across Europe? *European Journal of Human Genetics*, 14, 245-248.

Levitt, M. and Weldon, S. 2005. A well placed trust?: Public perceptions of the governance of DNA databases. *Critical Public Health*, 15(4), 311-321.

McCarthy, M.I., Abecasis, G.R., Cardon, L.R., Goldstein, D.B., Little, J., Ioannidis, J.P.A. and Hirschhorn, J.N. 2008. Genome-wide association studies for complex traits: Consensus, uncertainty and challenges. *Nature Reviews Genetics*, 9, 356-369.

Stranger, M., Chalmers, D. and Nicols, D. 2005. Capital, trust and consultation: Databanks and regulation in Australia. *Critical Public Health*, 15(4), 349-358.

Tutton, R. and Corrigan, O. eds, 2004. *Genetic Databases: Socio-ethical Issues in the Collection and Use of DNA*. London: Routledge.

PART 1
Benefit Sharing

Chapter 1

What Benefit Sharing Arrangements do People Want from Biobanks? A Survey of Public Opinion in Australia

Dianne Nicol and Christine Critchley

Public trust is a fundamental cornerstone in biomedical science in general and biobanking in particular (Campbell 2007). Essentially, in the biobanking context, members of the public who participate are being asked to put their trust in biobank operators and researchers to manage and keep secure their information and tissue, and to use them in ways that maximize any anticipated social and/or personal benefits. Clearly then, public trust and public participation go hand in hand. Members of the public are only likely to be willing to provide their tissue and information to biobanks if they have trust in the operators and users of biobanks and in the governance structures that regulate them (Kaye and Martin 2000, Chalmers and Nicol 2004).

A number of studies have attempted to measure the level of public trust in science in general, and in biomedical science in particular. Although a UK Parliamentary inquiry reported that there was a 'crisis in trust' in science (UK Parliament 2000, Stranger, Chalmers and Nicol 2005), other evidence suggests that trust in publicly funded biomedical research remains high (admittedly, though, there seem to be much lower levels of public trust in agricultural biotechnology. See, for example, Farquharson and Critchley 2004, Cormick 2007). Despite these positive views with regard to publicly funded biomedical research, the evidence also suggests that the level of trust decreases significantly when there is commercial involvement (Critchley 2008).

This chapter reports on a survey of public opinion relating to biobanking in Australia. The survey focused on the importance of the various 'benefits' that might emerge from biobanking and the ways in which the public thought they should be shared. Although there is an expanding international literature on benefit sharing in the context of biobanking (for example, Newfoundland and Labrador Department of Health and Community Services 2003, Irish Council on Bioethics 2005, Knoppers and Sheremata 2003, Laurie and Hunter 2004, Webster et al. 2008), scant attention has been paid to this issue to date in the Australian context.

There is no large-scale population biobank in Australia, nor is one planned in the near future. However, there are a number of National Health and Medical Research Council (NHMRC) funded 'biobank' collections, with specific biobank projects in Western Australia and Tasmania. There are also large numbers of tissue collections and small-scale genetic research databases. The NHMRC recognizes the importance of national uniformity in the development of regulatory and governance frameworks for biobanks. In particular, it is in the process of developing an information paper to facilitate best practices for the management and governance of biobank and tissue bank facilities, specific to an Australian research environment (NHMRC 2008). It is argued here that the formulation of best practice guidelines must take place against the backdrop of public consultation.

A Role for Public Consultation

It is increasingly being recognized that the development of regulatory and governance frameworks, including models for benefit sharing, cannot be achieved by policy makers in a vacuum, but must involve consultation with all interested parties. In particular, there is an increasing trend towards public consultation in the early, or upstream, phase of biobank establishment (Haddow et al. 2008). This process of upstream consultation is a necessary component of the information-gathering exercise, as it assists in understanding the motivations of members of the public to participate in biobanking and how the regulatory and governance frameworks must be shaped to satisfy their needs and allay their concerns. But it is more than this: it is an essential component of the democratic process (Longstaff and Burgess 2008).

There is now a body of evidence from consultations with various publics to show that many of the concerns relating to biomedical research in general, and biobanking in particular, coincide with the theoretical concerns canvassed in the academic and policy literature. The literature review provided in the report by Webster et al. (2008: 54-69), as well as the report itself, provides a useful overview of the outcomes of a number of those consultations. As might be expected, consent, privacy and access by third parties are all legitimate sources of concern. There is also an expected unease about commercial involvement in biobanking and biomedical research.

Yet there is evidence to suggest that people are quite capable of understanding why commercial involvement is necessary in order to bring the products of research into the pharmacy and clinic; and that they are still willing to participate provided that appropriate safeguards are in place to balance the costs and benefits of commercialization (for example, People, Science and Policy Ltd. 2002: 20-21, Webster et al. 2008: 19, 26). As such, it is appropriate, indeed necessary and vital, to canvass the various options for benefit sharing that have been mooted in the policy and academic literature with members of the public to test if it is possible to reach any sort of consensus on the way forward.

Benefit Sharing

It is widely recognized that, in the past, people generally participated in biomedical research for purely altruistic reasons, because of the likely benefit that such research could bring to society as a whole (Laurie and Hunter 2004, Campbell 2007). On this basis, it could be argued that the mere promise of new healthcare developments is sufficient to satisfy any benefit-sharing obligations. People may also have other personal reasons for participating, for example if they, or their families, suffer from a particular medical condition, which they believe may be alleviated as a result of the research.

There is evidence of increasing commercialization of biobanking itself and research using biobank resources, even where these are conducted in publicly-funded institutions (Merz 2003, Anderlik 2003). The usual justification is that commercial involvement is necessary to ensure that research results are actually translated into new developments in healthcare. In return, commercial parties require patent protection together with other commercialization guarantees in order to recover research costs, and to safeguard future profits. It is well recognized that the process of drug development is long, expensive and risky, but once products have reached the market there is good opportunity to profit.

From the public benefit perspective, commercialization seems to makes good sense, at least in theory: researchers get to do their research; investment by commercial parties is protected, which encourages them to develop new healthcare products; and the public interest is served through access to these new products. But, where commercial entities are involved, the people who are supplying the essential raw materials for purely altruistic reasons may feel marginalized and exploited (Andrews and Nelkin 1998). Capps et al. (2008: 29) emphasize the importance of respecting participant consent and the trust relationship, concluding that transactions between biobanks and industry ‘should be conducted above reproach in line with the stated public interests’.

Benefit sharing is being proposed both as an appropriate means of balancing the conflicting interests involved in biobanking and also as a potential solution to the problem of loss of public trust (Nicol 2006). Scholars have struggled to find a clear and cohesive normative foundation for benefit sharing (Chadwick and Berg 2001, Berg 2001, Weijer 2000, Gold and Caulfield 2003). But others have argued that trust of itself is sufficient rationale for benefit sharing (Laurie and Hunter 2004). In using the language of benefit sharing, it is important to acknowledge that financial benefit is not the only benefit that should be considered. The umbrella of benefit sharing should also extend out more broadly to include displays of altruism and gestures of benevolence.

The Present Study

While there has already been some public consultation work on biobanking in Australia (Williams 2005, Bruce 2006, Fleming 2007) to date there has been no nation-wide survey that is representative of the views of the Australian population. The need for comparative analyses of public opinion between jurisdictions also justifies the current study. Hence, this study provides a survey of public opinion that will be useful in informing future public consultations and the development of Australian regulatory and governance frameworks, as well as contributing to the international debate on biobanking and benefit sharing.

The survey instrument as a whole was split into three parts that were presented to respondents in the following order: (1) Biobanks in general; (2) Benefit sharing; and (3) Demographic characteristics. A full transcript of the survey can be obtained from the authors. This chapter reports on the results of part two of the survey, that is, the questions relating to benefit sharing. A full report of the results of the survey is in preparation.

The survey canvassed public opinion on both financial benefits, in the form of payment to donors or to the biobank for use, and other forms of benefit. A distinction was drawn between those arrangements that benefit the individual participant and/or their immediate family or community group and those arrangements that benefit society more generally. A further distinction was made between benefits that might flow to the Australian public and those that flow to people who are in need of humanitarian assistance in other countries.

Method

Procedure

A total of 1,000 Australians over the age of 18 years participated in a Computer Assisted Telephone Interview (CATI) that conformed to the ethical standards of the Australian Psychological Society. A sample size of 1,000 was deemed necessary to achieve an acceptable margin of error in responses (i.e., 3.1) and a confidence interval of 95 per cent assuming a 50 per cent split on each question (American Association for Public Opinion Research [AAPOR], 2007).

Telephone numbers were randomly selected from the electronic white pages to represent the proportion of residents residing in each Australian State and Territory. From these, 10,682 numbers were randomly selected by the CATI software for use in the survey. A quota was used to ensure that the sample of 1,000 was representative of each State and Territory. Of the 10,682 phone calls, 1,000 completed interviews were obtained, resulting in a minimum response rate of 19.0 per cent (maximum = 22.9 per cent) (AAPOR 2004).

Questions on payment to donors

The first series of questions assessed attitudes toward the payment of biobank donors. Respondents were asked, 'Using a scale from zero to 10, where zero means 'disagree strongly'; 5 (the midpoint) means 'mixed feelings'; and 10 means 'agree strongly', please indicate the extent to which you agree or disagree with the following:

1. Donors should receive payment from the biobank if their contribution is used for commercial company research.
2. Donors should receive payment from the biobank if their contribution is used for publicly-funded research.
3. Donors should receive payment from the biobank, even if their contribution is not used in research.

The first two questions in this series were randomized, but the third question was always presented last.

Questions on payment to biobanks

The next series of questions was designed to assess the public's attitude towards government funded biobanks receiving payment, and whether or not attitudes differ across the public and private, and Australian and international contexts. Respondents were first asked, 'Whether or not you believe that government funded biobanks should receive some payment for allowing other organizations to access their biobank resources'. The first four statements were randomized, while the fifth was always presented last. Respondents were asked to use the same zero to 10 scale as above. The statements relating to payment of biobanks were as follows:

1. The biobank should receive payment if their resources are to be used by an Australian commercial company.
2. The biobank should receive payment if their resources are to be used by an international commercial company.
3. The biobank should receive payment if their resources are to be used by a publicly funded Australian research organization.
4. The biobank should receive payment if their resources are to used by a publicly funded international research organization.
5. Biobank resources should be made freely available to all legitimate research organizations.

Questions on conditions on access

Attitudes towards access to Australian public biobank resources by commercial companies was assessed by three questions that followed the introduction, 'Before being allowed to access biobank resources, companies must …'. The questions were designed to specifically assess respondents' attitudes as to whether companies should be required to distribute any new treatments developed from public biobank resources at affordable rates to three groups: donors, developing countries and all Australians. A fourth question was designed to assess attitudes toward biobanks determining the cost of any treatments developed. Specifically the four questions were:

1. Companies must agree to make new treatments developed from biobank resources affordable to all those donors whose resources were used.
2. Companies must agree to make new treatments developed from biobank resources affordable to developing countries.
3. Companies must agree to make new treatments developed from biobank resources affordable to all Australians.
4. Biobanks should not be allowed to determine the cost of any new treatments developed by a company.

The same 10-point, disagree/agree scale was used, and the first three statements were randomized, with the fourth always asked last.

Questions on conditions on use of research results

The next series of questions related to attitudes towards the use of research results by other researchers using biobank resources. Respondents were asked for their 'thoughts on biobanks placing conditions on the use of its resources by other legitimate researchers'. Using the same 10-point agree/disagree scale, they were asked for their level of agreement to two statements. The first was, 'Before being allowed to access biobank resources, researchers must agree to give the biobank all of their research results that were found by using the biobank's resources. By research results I mean original data files, records and reports'. The second was, 'Before being allowed to access biobank resources, researchers must agree to publicly release all of their results that were found using the biobank's resources. By research results I mean original data files, records and reports'.

Questions on perceived importance of benefit sharing

The final set of benefit sharing questions was designed to assess the importance Australians place on benefit sharing in general. Two questions were asked after the following statement was read.

> The last set of questions all related to the issue of benefit sharing. That is, the sharing of any new discoveries or treatments that come from the use of biobank resources. Imagine again that you are approached by a genetic researcher from a government funded Australian university biobank. They ask for your permission to use a blood sample that was taken from you in the past for a medical procedure.

The first question asked respondents, 'When deciding on whether or not to participate, how important would it be to you to know about any benefit sharing arrangements the biobank has with other legitimate researchers or research organizations?' and the second, 'How much do you generally care about Australian university biobanks placing conditions upon the use of its resources by other researchers?' The response options ranged from 0 = 'not at all important', to 10 = 'very important'.

Results

In comparison to the Australian Bureau of Statistics (ABS) Census data for 2006, the sample was representative of the Australian population in terms of state, education, and occupation (i.e., no significant differences were found between the sample and ABS distributions), but was over represented by older individuals (M = 53.31 years; SD = 16.73; Range = 18-94) and females (66 per cent). The data was therefore weighted according to ABS (2006) proportions for gender and age groups (that is, 19-24 years;[1] 25-34 years; 35-44 years; 45-54 years; 55-64 years; 65-74 years; 75-84 years and over 85 years).

Figures 1.1 to 1.5 display the mean scores for each response to all questions in the benefit sharing part of the survey. The bars in each figure represent the range where the Australian population is estimated to lie (that is, 95 per cent confidence intervals). All comparisons across contexts were computed via SPSS (Version 16) using a one-way repeated measures (ANOVA) with difference contrasts. A difference was considered statistically significant it was less than $p<.001$.

Payment to donor

Figure 1.1 shows the mean level of agreement to the proposition that donors should receive payment for the three types of research use outlined above in the methods section.

The results in Figure 1.1 clearly illustrate that there is a low level of expectation that donors should be paid when the resources that they contribute to the biobank are used for research purposes, particularly when the research is publicly funded or is carried out by the biobank itself. All mean scores are below the mid point for all situations, with the exception of when the contribution to the biobank is used for commercial research. Here the upper limit of the confidence intervals reaches the mid point, suggesting that Australians are more inclined to accept the idea that donors should be paid if their resources are to be used by commercial researchers, but are still not fully convinced that this is appropriate. Mean agreement scores differed significantly across all categories, with higher agreement with the idea that donors should be paid if their contribution is used for commercial compared

1 One 18 year old was included in this group for weighting purposes.

to public research. Respondents were also less likely to agree that they should be paid when their contributions are used by the biobank itself in its own research compared with when their contributions are used by other public researchers.

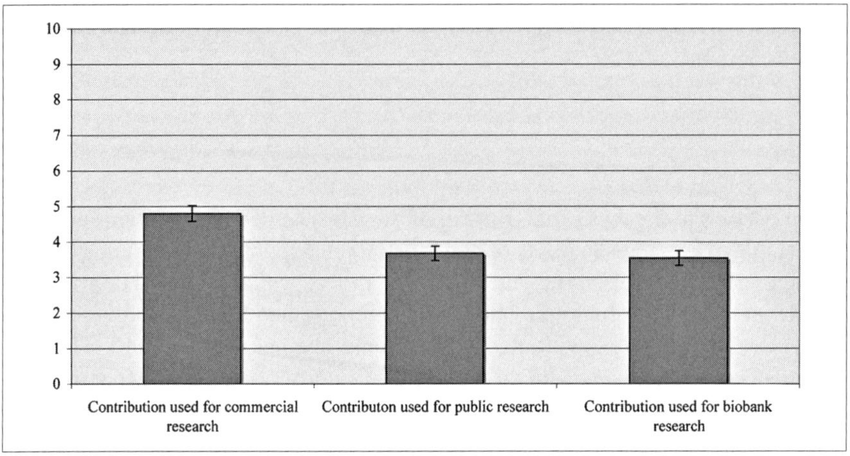

Figure 1.1 Receipt of payment by donor
Note: For Figures 1.1-1.4, where 0 = strongly disagree, 10 = strongly agree.

Access to resources

Figure 1.2 shows the mean level of agreement to the proposition that biobanks should receive payment for allowing access to their resources to the four types of organizations outlined above in the methods section.

These results indicate that, unlike payment to donors themselves, respondents have much higher expectations that biobanks should receive some form of payment for providing access for research purposes. As with other aspects of this survey, respondents draw a distinction between public and commercial research and Australian and international organizations. The results suggest that Australians are much more likely to agree that biobanks should be paid if their resources are used by commercial compared to public research organizations, and are also more likely to agree that biobanks should be paid if their resources are used by international compared to Australian organizations. These results further illustrate that the type of organization (public *v.* commercial) rather than its location (Australian *v.* international) is a more important determinant of expectations with regard to payment for access.

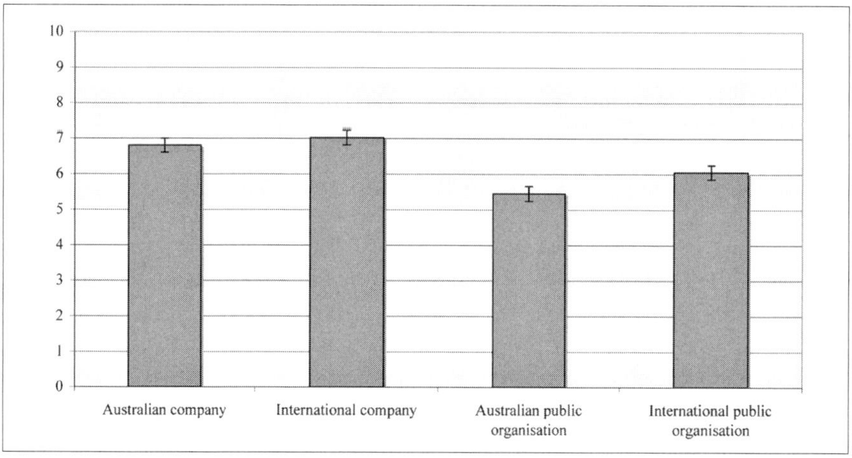

Figure 1.2 Payment by biobank for access to resources

Treatments

Figure 1.3 illustrates respondent views as to the types of conditions that should be placed on companies that access biobank resources for the development of new drugs and treatments. This set of questions specifically focused on the types of conditions respondents would like to see in terms of access to healthcare developments.

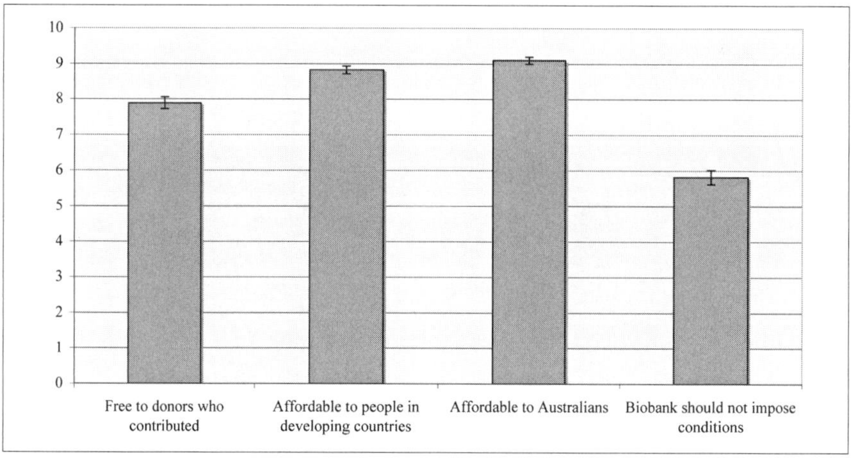

Figure 1.3 Conditions on access where companies make new treatments using biobank resources

Figure 1.3 suggests that Australians are generally strongly in favour of imposing conditions with regard to access to ensure that new treatments reach the community in a way that is affordable to them. General sharing of the benefits across the Australian community was slightly more important than making benefits more affordable to people in developing countries, and making benefits free to donors was viewed as less important than making them affordable to developing countries.

Respondents in this survey therefore seem to take a benevolent view with regard to access for people in developing countries. There also appears to be an expectation of specific benefit sharing with regard to the question of whether biobank participants should be provided with free access to healthcare benefits. Surprisingly, the mean of 5.8 with regard to the statement that biobanks should not impose conditions on access suggests that respondents are somewhat neutral on this point. Instead, we expected that respondents would express negative or strongly negative views on this point if they simultaneously expected benefits to be made affordable to others (that is, that they would want biobanks to impose conditions). The validity of this question is somewhat questionable, since, as the statement was framed negatively, it may have caused some confusion to respondents.

Research results

Figure 1.4 displays mean agreement scores relating to respondent views as to the types of conditions that should be placed on researchers who wish to access biobank resources for legitimate research purposes. These questions specifically focused on the types of conditions respondents would like to see in terms of release of research results.

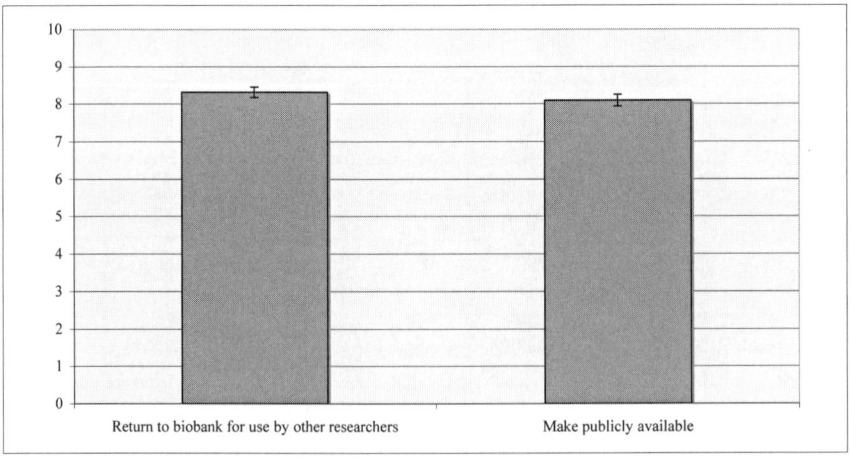

Figure 1.4 Conditions on access relating to the release of research results

Figure 1.4 suggests that Australians are strongly in favour of imposing conditions with regard to release of research results. They appear to favour both return of research results to the biobank and public release of research results. There was a slight difference in that returning research results to the biobank was marginally viewed more favourably than making results public.

Benefit sharing intention and care

Figure 1.5 displays mean importance scores relating to whether respondents wish to know about benefit sharing arrangements between biobanks and other legitimate researchers and research organizations. It also shows the extent to which they care about what these arrangements actually are.

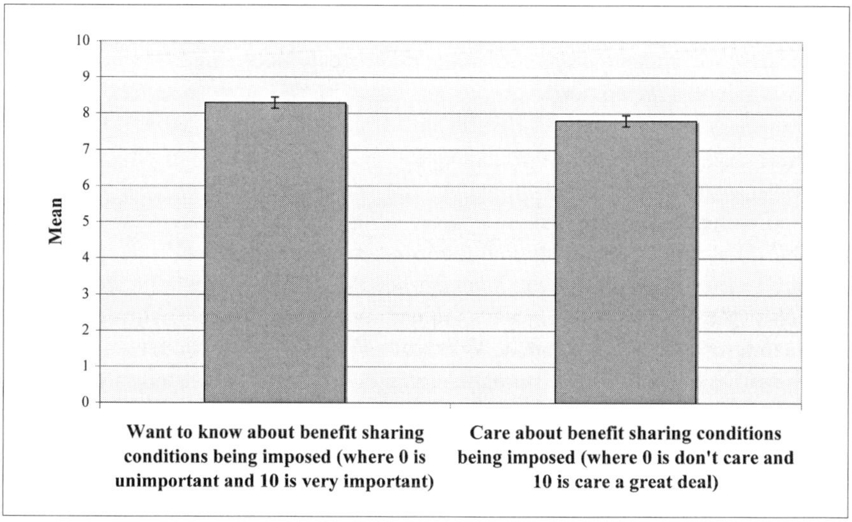

Figure 1.5 'Respondents' wish to know about benefit sharing arrangements

The results in Figure 1.5 indicate that respondents are highly in favour of knowing what benefit sharing arrangements are in place before deciding on whether or not to participate in biobank research. They also suggest that Australians do care about whether or not biobanks place benefit sharing conditions upon the use of their resources by other researchers.

Conclusions

These results illustrate that Australians tend to take an altruistic approach to their participation in biobank research. They also suggest that there is an expectation amongst the Australian community that other biobank donors should also participate for benevolent reasons. However, Australians do seem to have expectations that there will be some *quid pro quo* for their voluntary participation, not necessarily or just in the financial sense. There is an expectation that there should be public benefits from research using biobank resources, which are that research results should be shared and new healthcare products should be made widely available.

As a starting point, it appears that respondents do not see payment for participation as a desirable form of benefit sharing, although they may see this as being somewhat more justified in the commercial context. This is not to say that all forms of specific participant benefit sharing should be rejected. Indeed, there is a high level of support for benefit sharing in the form of free access to new healthcare developments for those who participate in biobanking.

Respondents reacted positively to questions directed at more general forms of community benefit sharing, in both the research and healthcare contexts. There is a high level of support for keeping research open, both by returning results to the biobank and by public dissemination. There is a similar high level of support for keeping healthcare products developed using biobank resources affordable to both the Australian community and the broader global community. There is also support for the notion that researcher users should pay the biobank for access to its resources, particularly if those users are commercial in nature. A similar finding was reported in the UK Biobank CATI survey (Webster et al. 2008).

The survey reported in this chapter provides valuable information on public attitudes towards benefit sharing in biobanking and the need for appropriate governance structures for benefit sharing, together with the types of benefit sharing obligations that should be included in best practice governance mechanisms. One of the most important findings from this study is that people care about benefit sharing and want to know what benefit sharing arrangements are in place before deciding to participate in biobank research.

These results lend support to the argument that, as a matter of course, benefit sharing should be explicitly addressed in biobank governance frameworks. If benefit sharing is ignored, the risk of reduced public support, especially for biobanks with industry links, will increase. Arguably, biobanks can ill afford this risk, as their success ultimately rests on public willingness to participate. Trust in operators, and an expectation that biomedical science will produce benefits for all Australians and for the broader global community are critical.

Acknowledgements

This research was supported by Australian Research Council Discovery Project DP0559760. The authors would like to thank Professor Margaret Otlowski and Dr Mark Stranger for assistance with survey design, the staff at the Swinburne University CATI facility for their efforts in conducting this survey, and to all those respondents who kindly gave their time to be interviewed. Thanks are also due to Gordana Bruce for assisting with data collection.

Bibliography

American Association for Public Opinion Research. 2007. *Margin of Sampling Error*. [Online] Available at: http://www.aapor.org/marginofsamplingerror?s=margin%20of%20error [accessed 22 February 2009].

Anderlik, M.A. 2003. Commercial biobanks and genetic research: Ethical and legal issues. *American Journal of Pharmacogenomics*, 3(3), 203-215.

Andrews, L. and Nelkin, D. 1998. Homo economicus: Commercialization of body tissue in the age of biotechnology. *Hastings Center Report*, 28, 30-39.

Berg, K. 2001. The ethics of benefit sharing. *Clinical Genetics*, 59, 240-243.

Bruce, D. 2006. *Joondalup Family Health Study: Report on the Results from the Initial Community Survey and from the Community Workshop Deliberative Survey November/December 2005.* Prepared for Western Australian Institute for Medical Research, University of Western Australia. [Online] Available at: http://www.jfhs.org.au/ [accessed 27 February 2009].

Campbell, A.V. 2007. The ethical challenges of genetic databases: Safeguarding altruism and trust. *Kings Law Journal*, 18, 227-245.

Capps, B., Campbell, A.V. and ter Meulen, R. 2008. *Access to the UK Biobank Resource: Concepts of the Public Interest and the Public Good.* Report Commissioned by the UK Biobank Ethics and Governance Council. [Online] Available at: http://www.egcukbiobank.org.uk/assets/wtx048966.pdf [accessed 27 February 2009].

Chadwick, R. and Berg, K. 2001. Solidarity and equity: New ethical frameworks for genetic databases. *Nature Reviews Genetics*, 2, 318-321.

Chalmers, D. and Nicol, D. 2004. Commercialization of biotechnology: Public trust and research. *International Journal of Biotechnology*, 6, 116-133.

Cormick, C. 2007. What does the public really think and who do they really trust?, in *Human Biotechnology and Public Trust: Trends Perceptions and Regulation*, edited by M. Stranger. Hobart: Centre for Law and Genetics Occasional Paper No. 7, 115-129.

Critchley, C. 2008. Public opinion and trust in scientists: The role of the research context, and the perceived motivation of stem cell researchers. *Public Understanding of Science*, 17, 309-327.

Farquharson, K. and Critchley, C. 2004. Risk, trust and cutting-edge technologies: A study of Australian attitudes. *Australian Journal of Emerging Technologies and Society*, 2, 121-146.

Fleming, J. 2007. Issues with tissues: Perspectives of tissue bank donors and the public towards biobanks and related genetic research, in *Human Biotechnology and Public Trust: Trends Perceptions and Regulation*, edited by M. Stranger. Hobart: Centre for Law and Genetics Occasional Paper No. 7, 184-201.

Gold, E.R. and Caulfield, T.A. 2003. *Human Genetic Inventions, Patenting and Human Rights*. [Online] Available at: http://www.cipp.mcgill.ca/en/publications/author/?q=E.+Richard+Gold [accessed 27 February 2009].

Haddow, G., Cunningham-Burley, S., Bruce, A. and Parry, S. 2008. Generation Scotland: Consulting publics and specialists at an early stage in a genetic database's development. *Critical Public Health*, 18, 139-149.

Irish Council for Bioethics. 2005. *Human Biological Material: Recommendations for Collection, Use and Storage in Research*. [Online] Available at: http://www.bioethics.ie/index.php/human-biological-material-for-research [accessed 27 February 2009].

Kaye, J. and Martin, P. 2000. Safeguards for research using large scale DNA collections. *British Medical Journal*, 321, 1146-1149.

Knoppers, B. and Sheremata, L. 2003. Beyond the rhetoric: Population genetics and benefit-sharing. *Health Law Journal*, 11, 89-117.

Laurie, G. and Hunter, K. 2004. Benefit-sharing and public trust in genetic research, in *Blood & Data: Ethical, Legal and Social Aspects of Human Genetic Databases*, edited by G. Árnason, S. Nordal and V. Árnason. Reykjavik: University of Iceland Press and Centre for Ethics, 323-331.

Longstaff, H. and Burgess, M. 2008. Recruiting for representation in public deliberations on the ethics of biobanks. *Health Law Review*, 16, 12-15.

Merz, J.F. 2003. On the intersection of privacy, consent, commerce and genetic research, in *Populations and Genetics: Legal and Socio-Ethical Perspectives*, edited by B.M. Knoppers. Netherlands: Koninklijke Brill NV, 257-268.

Newfoundland and Labrador Department of Health and Community Services. 2003. *Policy Implications of Commercial Genetic Research in Newfoundland and Labrador*. [Online] Available at: www.nlcahr.mun.ca/research/reports_search/DP_Final__Report.pdf [accessed 22 February 2009].

NHMRC. 2008. Development of an NHMRC information paper on biobanks and tissue banks, in *NHMRC Tracker April 22, 2008*. [Online] Available at: http://www.nhmrc.gov.au/media/tracker/tracker08/080422.htm [accessed 28 February 2009].

Nicol, D. 2006. Public trust, intellectual property and human genetic databanks: The need to take benefit sharing seriously. *Journal of International Biotechnology Law*, 3, 89-103.

People Science & Policy Ltd. 2002. *A Question of Trust: A Consultation Exploring and Addressing Questions of Public Trust*. Report prepared for the Medical

Research Council and Wellcome Trust. [Online] Available at: www.ukbiobank.ac.uk/docs/people-science-policy.pdf [accessed 27 February 2009].

Stranger, M., Chalmers, D. and Nicol, D. 2005. Capital, trust and consultation: Databanks and regulation in Australia. *Critical Public Health*, 15, 349-358.

UK Parliament. 2000. *Third Report: Science and Society: Summary*. [Online: House of Lords Select Committee on Science and Technology] Available at: http://www.publications.parliament.uk/pa/ld199900/ldselect/ldsctech/38/3802.htm [accessed 27 February 2009].

Webster, A., Brown, B., Douglas, C., Lewis, G., Kaye, J., Tutton, R. and Williams, N. 2008. *Public Attitudes to Third Party Access and Benefit Sharing: Their Application to UK Biobank.* Final report to the UK Biobank Ethics and Governance Council. [Online: York, Science and Technology Studies, University of York] Available at: http://www.egcukbiobank.org.uk/meetingsandreports/index.html [accessed 27 February 2009].

Weijer, C. Benefit-sharing and other protections for communities in genetic research. *Clinical Genetics*, 58, 367-368.

Williams, C. 2005. Australian attitudes to DNA sample banks and genetic screening. *Current Medical Research and Opinion*, 21, 1773-1775.

Chapter 2
Reconsidering Altruism, Introducing Reciprocity and Empowerment in the Governance of Biobanks

Nadja Kanellopoulou

This work addresses the disparity in the balance of legal power between researchers and research participants in current approaches to biobanking, and the associated problems this creates for the regulation of their interaction. It examines enduring regulatory ambiguities in UK policies and highlights that while these policies remain linked to altruism, recent practices in the evolution of biobank governance indicate a strong interest for collaborative models that redistribute some of that power to research participants. This chapter suggests that there is an urgent need to clarify fundamental principles and assumptions in regulating biobanking research relationships, which cannot be achieved without reviewing the shortcomings of the donation model.

The original analysis contained in the first part exposes the inconsistencies in the direct endorsement of the unconditional gift (donation) relationship idiom in research. The work discusses that the engagement between biobanking participants and researchers is a long-lasting affair and it is important for law to review its nature more carefully. In this engagement, cooperation and trust between participants and researchers are paramount. This chapter proposes a new and better way for conceptualizing their interaction by using *reciprocity*. In developing ways to apply rules of reciprocity in biobanking governance, it puts forward a new research ethics principle, *empowerment*, to establish a normative basis and a set of criteria for reciprocation.

On the basis of empowerment, new legal models can be developed to effectively regulate the relationship between biobanking participants and researchers. This work proposes a new model under which research 'donations' are not defined as 'free gifts' but instead as conditional and ongoing returns of favours between participants and researchers, or, *conditional gifts*. The analysis in this chapter is inspired by legal, anthropological, ethical and economic investigations of collaborative interactions and collective endeavours. It seeks to introduce a novel way for conceptualizing biobanking research relationships.

Regulatory Appeal for Altruism

Definition of altruism

There is extensive use of altruism in the regulation of genetic and genomic research in the UK to promote charitable and beneficent practices towards biomedical research. Long, theoretical debates exist about the meaning of altruism.[1] In this chapter, I refer to altruism as 'action in the interests of another or the disposition to act in the interests of another' (Birks 2000: 14). I examine current guidance on the use of human biological samples in research in order to highlight a persistent and particular *enclosure* of potential participants' motivations in research. A number of reports have been commissioned in the UK since the mid-1990s to address legal and ethical issues in the use of human tissue samples for research purposes. These advisory bodies frame the relation between potential participants and research as solely altruistic, in various ways, and name the removal of tissue for research purposes as a 'genetic gift'.

For example, in its 1995 report, the Nuffield Council on Bioethics proposed that:

> … [t]he act of providing tissue … clearly indicates what is involved is a *gift* … [which is] … *free of all claims* … (Nuffield Council on Bioethics 1995: 68, my emphasis)

According to the report, the genetic gift would be considered as:

> … a *gratuitous voluntary* transfer from the true owner in possession with *no expectation of its return* … (Nuffield Council on Bioethics 1995: 68, my emphasis)

In their 2001 report, the Medical Research Council Working Group on Human Tissue and Biological Samples reiterated that position and recommended that:

> … tissue samples donated for research be treated as *gifts* or *donations* … (MRC 2001: para 2.2, my emphasis)

and further suggested that:

---

1 The concept has a long and complex history dating back to the 1850s. Debates about altruism have attracted a great deal of attention from a wide range of disciplines as diverse as sociology, social psychology, neurology, behavioural biology, philosophy, game theory and economics. A large part of these debates lies beyond the scope of this chapter but it is important to be aware of their multiplicity, as the concept of altruism is used in different and independent ways across disciplines. For more discussion see (Kanellopoulou 2008).

> … [t]his is preferable from a moral and ethical point of view, as it promotes the "*gift relationship*" between research participants and scientists, and underlines the *altruistic motivation* for participation in research … (MRC 2001: para 2.2, my emphasis)

The term that the MRC Working Group refers to as 'gift relationship' was originally used in social research studies investigating patterns of exchange in primitive societies (Mauss 1990). In the 1960s, these studies were used by scholars for the elaboration of theories and policies on voluntary blood donation (Titmuss 1970). They influenced the formulation of policies which paved the way for developing voluntary models for organ donation in the UK (Waldby and Mitchell 2006). According to medico-sociological literature, the altruistic UK model is not universal and different approaches have been developed in other countries, where two systems of altruistic and paid donation exist in parallel, as in the US (Katz 2006, Mason and Laurie 2006, Oberman 2006, Waldby and Mitchell 2006).[2]

The rationale of policies introduced in the UK in the 1960s favoured the use of altruism in the regulation of blood donation. The same rationales are now being used to promote public provision of tissue for genetic research (Shaw 2008). Recent critics highlight that these attempts fail to understand that 1960s models were underpinned by particular ideas on the welfare state within a specific socio-political climate, which conflate unconditional gift (donation) with a relationship that should be understood as part of a chain of exchanges that encapsulate giving, receiving and reciprocating (Hayden 2007, Shaw 2008, Tutton 2004).

Careful interpretation of this literature allows for multiple meanings of how gifts can be conceptualized and leads to differing conclusions as to whether gifts are in reality 'free'. In the research context in particular, some scholars comment that the 'traditional' (unconditional) gift model is not altogether successful; it needs to be supplemented with an element of *mutual obligation* for all parties involved (Gottlieb 1998, Kanellopoulou 2004, Levitt and Weldon 2005). I support the argument that the importation of the early discourses about donation in the context of biobanking is misconstrued. The relevant bodies mistakenly conceptualize the transfer of DNA tissue for research as a one-way, altruistic transaction. One could argue further that this is aimed chiefly at facilitating the unproblematic, free supply of biological material for research purposes and not necessarily at protecting participants' interests in the use of human tissue in research. For example, Laurie (2002: 313) comments that '[i]n practice, [the continuing use of the unconditional gift model] has *considerable utility for the recipient* … [as] [it] provides broad scope for the future use or disposal of the gift' (my emphasis).

2 Healy (2006) helpfully suggests that successful systems rest on the *fairness of the exchange* rather than the purity of a donor's altruism or the size of a financial incentive.

The Problem of Unequal Relationships

A range of problems beleaguer the direct endorsement of the donation model in research. These problems are owed to two unfortunate assumptions, founded in the surrender provision and the unquestioned use of altruism in research.

The surrender provision

According to existing UK guidance, the legal notion of genetic gift would be considered as a free voluntary transfer with no expectation of its return. Adding to that notion, the Nuffield Council on Bioethics advocates that this 'gift' involves: '... [the] surrender of all interests ... in the removed tissue on the part of the person from whom it was removed' (Nuffield Council on Bioethics 1995: 68). The report further assumes that the person from whom the tissue is removed does not have: '... the slightest interest in making a claim to [the donated sample] once it is removed ...' (Nuffield Council on Bioethics 1995: 68). I call this the 'surrender provision'. Reinforced by the recommendations of the MRC Working Group, it essentially stipulates that any proprietary rights the donor might have in their tissue should be transferred with the control over the use of the tissue to the recipient of the gift (MRC 2001: para 2.2)! This position aims to tie with the European Convention on Human Rights and Biomedicine (1997) position that 'the human body and its parts shall not, as such, give rise to *financial* gain'.[3] In the MRC 2001 report, the Working Group advises that:

> ... [they] fully support the principle: the *sale* of human biological samples for research is not ethically acceptable. Therefore, while reasonable expenses (e.g. travel expenses) may be reimbursed, research participants should never be offered any *financial or material* inducement to donate biological samples for research ... (MRC 2001: para 2.5)

In seeking to adopt the approach of the international document, the report casts off (financial) inducement and sale of samples on the part of the participants as ethically unacceptable gain. Interestingly, it also expands the notion of unacceptable gain from 'financial' to other 'material' gain although there is no such disclaimer

3 Article 21 prohibits commerce in the human body or its parts; Article 22 requires a regime of informed consent to be applied where, following a medical intervention, a human body part is stored or used for a purpose other than that for which it was removed. Beyleveld and Brownsword (2004: 88) helpfully point out that this position is associated with one or other of two underlying claims: either that humans have no proprietary rights in their own bodies or body parts (including body tissue and fluids); or that, insofar as humans do have such proprietary rights, they have no right to exploit the commercial value of their bodies or body parts.

anywhere else in the relevant articles of the Convention or of its Explanatory Report, which states that:

> … [u]nder this provision organs and tissues proper, including blood, should not be bought or sold or give rise to *financial* gain for the person from whom they have been removed or for a third party, whether an individual or a corporate entity such as, for example, a hospital … (Explanatory Report 1997: para 132)

The report mentions no other type of gain. By considering that any proprietary rights that the donor might have (in their tissue) should be transferred with the control over the use of the tissue to their recipient, the UK position seems to imply that property rights exist – even though there is an extensive antipathy to property in the body! The MRC Working Group does not discuss what these rights are or what their scope would be. This position is ambiguous and also ignores the significance of other, not necessarily financial (or material) interests of 'donors' in the samples. It is unclear and this is a problem, because when debating who can have control over the transfer and use of the samples, one needs to be able to ascertain what kinds of rights can be asserted on them and by whom. Scholars have criticized the current position on property as seriously incongruous (Laurie 2002, Mahoney 2000). I argue further that in practice, it is a way to circumvent claims for control and better balance in the power afforded to researchers and research participants. The question of whether participants can attain a better degree of power simply remains unaddressed in UK frameworks.

The surrender provision considers the value of tissue as limited and potential participants as having no interest in it (Tutton 2004, Waldby and Mitchell 2006). Alternatively, we could refer to this as the provision of 'relinquishment' or 'disinterestedness' in its dangerous hypothesis, if not presumption, that research participants do not have a stake in what happens to 'donated' samples once separated from their body (Carrier 1995: 166). It seems to rely on a second major assumption, that of altruism, which merits careful consideration, especially when combined with current regulatory language promoting notions of 'solidarity' and 'common interest' in research.

The reinvention of genetic solidarity

In 2002, the Human Genetics Commission proposed an interpretation of 'genetic solidarity', also warmly adopted by the UK Department of Health (2003: para 6.26):

> … we all share the same basic human genome, although there are individual variations which distinguish us from other people. Most of our genetic characteristics will be present in others. This sharing of our genetic constitution not only gives rise to *opportunities to help others* but it also highlights our *common interest in the fruits* of medically-based genetic research … (HGC 2002: 38)

In their reference to notions of *sharedness* of our genetic constitution, the HGC sought to emphasize a 'special moral relationship' and possibly a related obligation to research participation. The Commission promotes two key elements in support of such obligation: 'opportunities to help others' and a 'common interest in the fruits' of research. Critical questions about this position remain unanswered, for example, whether such a special moral relationship linking us to one another really exists; and, if it does, whether it ought to be constructed as a unidirectional giving to public and private entities for some long-term, intangible common good. Further, is there a common interest in the fruits of research that is 'shared by all', and if so, on what grounds and in what terms can this interest be properly defined, balanced and understood in the UK context? (Capps et al. 2008: 27-28).

By constructing genetic solidarity in such a way, the HGC reiterated the interaction between potential 'donors' and researchers as essentially a 'free-gift relationship' which primarily raises obligations on the part of the former. This is extremely problematic because both the gift metaphor and genetic solidarity proposition can be used to *enclose* the research relationship as a positive moral responsibility of potential donors towards the wider public and thereby ignore any broader claims on the part of 'donors' that may have some legitimate basis (Ratto and Beaulieu 2007: 184, Waldby and Mitchell 2006). Why must one give a tissue sample to research only as a free gift? One can argue that there is a comfortable social appeal in the use of charitable, 'feel-good' discourses and laudable acts. On a more practical note however, substantial utility considerations are involved in instituting free, gratuitous and therefore unconditional gifts, precisely in order to allow an unhindered and wide scope of possible future uses, without further interference. By appealing to discourses of moral obligation, gift relationships, solidarity and common interest, regulatory discourse encourages the exclusion of participants from retaining any interest in the fate and use of samples in research.

The rhetoric of altruism

Several criticisms have been expressed about the shortcomings of the use of genetic altruism in UK advisory guidance documents. I find the sole reliance on altruism in the research context unfortunate on at least four counts:[4]

- the erroneous hypothesis that the same circumstances apply in blood donation, organ transplantation and genetic research, which ignores the polysemic nature of gifts in these contexts (Haddow 2006, Healy 2004, Shaw 2008);
- the lack of evidence that research participants act solely for charitable reasons (Merz et al. 2002: 969);

4 As mentioned earlier, UK policy documents seem to define altruism as unselfish regard for the welfare of others. This approach promotes the provision of a benefit to others without expectation of payment, compensation or recognition.

- the failure to acknowledge other interests that become engaged, apart from any charitable ones (Waldby and Mitchell 2006);
- the risk of altruism becoming an exploitative device at the expense of participants' interests (Dickenson 2004).

These failings have grave implications for the future success of research with human tissue, to the extent that they are linked with the motivations and trust of research participants. Ultimately, such altruistic rhetoric is owed to either institutional fear of shackling medical progress, or a reluctance to recognize the interests of sources in research to be conducted with their tissue – or both. By assuming that altruism is the only reason why participants agree to participate in research and by stipulating the surrender of rights over samples, current guidance creates barriers to an equitable power balance between researchers and participants that limit the latter's interests (Andrews 2006). There are several normative and practical reasons why the transfer of human biological samples in research biobanking should not be considered automatically as a 'gift' free of all claims. These are explained directly below.

Participants' fears of being 'suckered': under the current altruistic model, participants are required to participate 'for nothing' yet contribute to private profit which furthers interests other than their own. Hank Greely (DNA Fingerprinting and Civil Liberties 2005) discusses this as a major emerging problem:

> … it's not so much an ethical obligation to share some of the profits, as it is a desire to avoid having people feel that they're suckers. You know, "I gave away something for free that you got awfully rich from. You took me for a ride. You mistreated me. I was a sucker", which, whether or not it's a strong ethical argument, doesn't have good political consequences for support of research …

There is emerging evidence of negative reactions to such lack of balance, encouraged by concern that participants may be exploited (Katz 2006). This can be defined as the *injustice in using one's gift to make another's profit* (PropEur Report 2006, HGC 2001). The importation of purely altruistic models imposes problematic obligations on participants to give unconditionally to research precisely because the *language of altruism implies no obligation on the part of the researchers and/or other profiteers* with respect to the donated material. Donna Dickenson (2004: 113) puts this as: '… altruism has too often tended to be one-way – from research populations to researchers – and one-way altruism is better called exploitation …'.

Other interests take precedence: a number of interests are not being taken into account by the current model, such as the interests of participants in influencing the use of samples for research purposes, for example, in obtaining information about illness in one's family or about one's descent or in ways not contrary to their belief-system (Wendler 2002). These can include reservations about commercialization of future research or the level of control participants may have on the direction of research (Sumner 2007, PropEur Report 2006, Haddow et al. 2004, HGC 2001). Concerns about participants' power and/or control in research are critical in developing collaborative mechanisms. They are linked both with ethical principles of fairness but also practical considerations on willingness to participate in research.

Principles of fairness: principles of distributive justice require that all legitimate stakeholders receive a 'fair share' of the benefits and burdens resulting from research (Pullman and Latus 2002). Ethical principles internationally support this claim; the fact that they are hard to pin down does not mean that the principles cannot be invoked (Knoppers 2000). This does not imply that it is currently possible for stakeholders to receive an *equal* share nor benefits need only be measured in monetary terms. For example, in benefit-sharing proposals, provisions are anticipated for groups to benefit through health care, hospital, and other community facility improvements (HUGO Ethics Committee 2000). Greely (2001: 227) argues further that 'fair' treatment of prospective research participants should require a discussion of whether the research involves, directly or indirectly, commercial interests; that they should be able to take that information into account when considering whether to participate; and, that commercial benefits should be shared 'in some manner' with the group who 'made the commercial benefits possible'.[5]

There is no absolute gift: the transfer involved in the genetic research gift is not an absolute one; scholars in very recent years note the right to withdraw from research at any time as acknowledged by current advisory guidance. For some, the withdrawal feature evinces a moral sensibility concerning the relation of participants to their extracted tissues (Winickoff 2007: 445) which arguably signals a major shift in research participation. The fact that the 'gift' is revocable means that it cannot be treated as an unconditional transfer. The possibility of withdrawal raises a critical difference between the paradigm of pure donation and genetic research. For some scholars, firm protection of withdrawal rights indicates a good faith attempt to endow participants with limited but real rights of control over (their) tissues (Gottlieb 1998: 190, Winickoff 2003: 218). I agree that the right to withdraw at any time from the research without having to explain why, and without penalty, preserves the voluntary nature of participation and awards participants with 'real' rights to some extent. Indeed the possibility of withdrawal

5 This view resonates in subsequent work by scholars who call for sharing through giving benefits to the collective and for rethinking both about the *process* and the *substance* of the relationship between researchers and research participants, as commercial interests in large-scale research biobanks resources are on the rise.

may imply some degree of recognition of participants' interests against abuse of (their) tissues yet this approach, when combined with one-off consent, falls short of empowering participants in having a say in the overall research endeavour (Laurie 2002: 312, McHale 2004: 85).

Lack of flexibility and interest in models for better control: the assumption of altruism denies the existence of any other rights or interests but for the charitable intentions of donors. For example, a conditional model would not prevent prospective participants from volunteering on the basis of charitable intentions only; it can include the possibility that the gift would be free of any conditions. This means that, even if one is to assume that prospective participants have only charitable intentions, under a conditional model, their rights and interests would be respected. This cannot be said in the case of the model of pure altruism which denies that any other interests exist.

The two assumptions of relinquishment and altruism lie at the heart of current discourses of disempowerment, put forward by scholars in recent years (Andrews 2004, Laurie 2002). They are also linked with debates on property rights in the body which could guarantee higher degrees of participant control in the use of samples in research (Hardcastle 2007). I contend that the critical question for the protection of participants' interests in this context is whether they can have a degree of better control of the *use* that is made of samples as well as other rightful claims, such as for the time, effort and resources that they contribute to research (MRC 2001: para 9.2, Bellivier and Noiville 2006: 99). A number of models have been put forward in recent years aimed at resolving the tension between disempowering rationale and collaborative approaches. These range from proposed property rights in various forms, patient advocacy agreements and benefit sharing proposals for local communities. These models are based either on: (1) 'do-it-yourself' attempts to secure better research oversight and proprietary control over the outcomes of research; for example, via advocacy contracts and calls for the recognition of indigenous property rights (Amani and Coombe 2005, Merz et al. 2002); or on (2) humanitarian approaches towards aiding or recompensing research participants in the form of benefit sharing proposals (Haddow et al. 2007, Pullman and Latus 2002). The former are inspired by a wish to establish proprietary legal rights for research participants over the use of their tissue. Bodily property rights are met by an entrenched antipathy both in jurisprudential and policy terms, yet in academic scholarship there is renewed interest in their viability (Hardcastle 2007, Mason and Laurie 2001). The latter need further clarification, both in regards to their normative basis as well as their practical aspects (Haddow et al. 2007, Hunter 2006, Nicol 2004). The collaborative models above have limitations in that they promise either excessive or rather limited degrees of power or control for research participants (Kanellopoulou 2008). I propose that a better approach for law to protect participants' interests would be to focus on the nature of their interaction with researchers and describe it as an ongoing cooperation and dynamic relationship with special obligations for both sides.

A Dynamic Engagement

Long-standing interdisciplinary research exists on understanding gift relationships. This literature draws on sociological and anthropological theories about the nature of gifts, exchanges and ensuing obligations, where gifts are studied as constitutive of positive social relations (Carrier 1991, Douglas 1990, Frow 1997, Mauss 1990, Osteen 2002, Titmuss 1970). Political economists who research contemporary trends in the changing economic value of human tissue highlight that in these systems of 'giving, receiving and reciprocating, gifts act more like *loans*' and that there exists a major re-evaluation of the 'seemingly mutually exclusive relationship' between gifts and commodities (Waldby and Mitchell 2006: 15 and 24).[6]

In these discourses, gifts and commodities are seen as polar terms that define a *continuum* along which one can place existing transactions, friendships or relationships (Carrier 1995: 190). These transformations of the status of human tissue in and out of waste, gift, and commodity forms are seen as types of circulating value (Appadurai 1986), involving processes that determine not only the economic value of things but also the context in which other symbolic and social meanings of things are developed (Lock 1997). The transformations are different phases in the change of status of human tissue, defined by the *relationships* in which tissue circulates at a particular time.[7] Current frameworks are not equipped to address the cultural implications of these status shifts and do not adjust protections accordingly. This lack of insight leads to misconceptions and difficulties. I contend that the biggest difficulty with regulating research participation in UK is the *fallacy of altruism* which prevents the reduction of continuing trust-related tensions between altruistic motivation to participate and sources' interests in how samples are used and in what kinds of research are being pursued (Chalmers and Nicol 2004, Sumner 2007).

In line with the above considerations, I suggest that the interaction between biobanking participants and researchers should not be viewed as just an exchange of samples but instead as a continuous, collaborative *engagement*. In defining the reasons and obligations derived from this interaction, altruistic or charitable intentions can form part of the reasons for engagement but these intentions are not necessarily the only ones, a fact which current regulation continues to ignore. This shift is paramount for the viability of relationships that require long-term participation and commitment on the part of participants. In order to take account

6 Similarly, Frow (1997: 110) comments that the '... transaction ... perhaps bears rather more resemblance to a *loan* than to an absolute gift or the alienation of a property right'.

7 This understanding is pervasive in anthropological scholarship since the mid-1980s on the circulation of gifts, commodities and related modes of production where culture is seen as a dynamic, shifting, contested terrain that is constantly shaped by – and shaping – a changing political and economic context (Appadurai 1986, Ferguson 1988, Komter 2005, Kopytoff 1986).

of intentions and interests other than charitable ones, a new conceptual framework is needed to institutionally acknowledge participants' interests in research as well as establish mechanisms to evaluate their contribution and enable the distribution of benefits from research. I propose that such a framework requires the clarification and establishment of rules of reciprocity.

Reciprocity and cooperation

Reciprocity imposes an obligation to return a favour or resource. In the highly influential piece on the function of reciprocity, Gouldner (1960: 171) says that the norm of reciprocity makes two demands: (1) that people should help those who have helped them; and (2) that people should not injure those who have helped them. According to these norms of reciprocity, the obligation to return a favour or resource does not necessarily require that the same favour or resource is returned. In many cases, the returned favour or resource can be a return in kind (especially where it is no longer possible, practical or desirable to return that favour or resource), as long as the return is assessed on a fair and mutual basis.

Reciprocity is linked to the idea of mutuality according to which obligations can be imposed only contingently in response to the benefits conferred by others (Gouldner 1960, Watkins 2006). *Mutuality* requires that parties are jointly bound as regards the benefits and risks from their interaction and that one's obligation of return is conditional to the value of the benefit received. *Reciprocity* not only requires that parties are jointly bound as regards the benefits and risks from their interaction but also that a *balanced* return exists between respective parties to shelter ongoing cooperation. There is a significant body of work from philosophical, economic, biological, anthropological and sociological perspectives on the function of reciprocity but very little reflection on this work from a legal point of view. The question I will address here is, how can we apply this expertise in law in order to achieve a better balance between the interests of research participants, researchers and research institutions? I contend that there exists a workable way to integrate such thinking in the development of legal frameworks by focusing on the *collaborative qualities* it involves. At a conceptual level, as the focus of biomedical research is shifting from individuals to populations, such integration is in line with current shifts in emphasis in bioethics, which highlight the need to respect diverse cultural values (HUGO Ethics Committee 2007). This shift in emphasis goes in pair with shifts within the scientific community. Here a new desire exists to invest in long-term and continuous relationships with research participants as population genomics is increasingly geared to include richer phenotypic and environmental data sets in the study of genotype.

In developing thinking to promote the collaborative qualities these emerging trends support, I see it necessary to take account of economic and psychological studies to highlight that reciprocity is achieved with *cooperation*. Cooperation happens when people believe that mutual synergy is generally a beneficial strategy to all, rather than individual action (Hayashi et al. 1999). This means that partners will collaborate when they think that their partner will cooperate. I see this as a situation where one party's expectation of another's cooperation is founded on two basic principles: (1) general *trust*; and (2) a sense of *control*. Trust is often enhanced by previous positive collaboration and experience between the parties. Trust-building helps achieve heightened levels of a sense of control but is not the only factor that influences one's sense of control (Kollock 1998). On the other hand, trust is at risk in negative accounts and histories of exploitation in research.

There exists a plethora of empirical evidence on the pivotal role of cooperation in establishing successful partnerships in several contexts, most notably in economics and psychology research (Hardin 1993, Hayashi et al. 1999, Kollock 1998). In genetic research, cooperation and communication has been achieved with extremely advantageous results by advocacy groups and patient groups formed to facilitate research in their particular area of interest, and to manage collective resources for this goal (Gitter 2004, Terry et al. 2007). The majority of these initiatives focused on ways to secure better collective influence in the direction and use of research and further distribution of research results to their community, achieved by using their *commonly owned* resources as leverage.

Healthy mechanisms of representation and communication can greatly enhance research participation rates, participants' trust and, by extension, project sustainability (Winickoff 2008). Attracting and retaining participants requires that participants trust that research is managed in a way consistent with their core values and expectations. I propose that cooperation and trust largely depend on a proper understanding of *reciprocity* which requires not only a shared basis for participants' and researchers' commitment but also that responses are balanced in order to guarantee ongoing cooperation, with the expectation that the parties respond to each other in similar ways.[8]

Gouldner (1960: 171) notes that reciprocity becomes a trust-building mechanism that helps stabilize and maintain social relationships.[9] In this respect, he successfully recognizes the complex balancing relationship between trust and control insofar as trust and control are both functionally necessary for the generation and maintenance of social order. In line with these considerations, the understanding of what constitutes appropriate reciprocation will depend on the observance of particular cultural and ethical norms and on the specifics of the dynamic between the parties. The reciprocal obligation need not be a mere 'tit-for-tat' since the gift recipient's means may not be adequate to return the full economic value of the giver's gift (Schmidtz 2005: 458).

8 An example that signals a move in this direction is the increasing calls for the development of provisions obliging researchers to disclose any potential commercial interests derived from the use of samples.

9 He considers the role of reciprocity as a tool to assist relevant parties in coping with the disruptive potentialities of power differences; for Gouldner, reciprocity helps achieve stability by circumscribing exploitative relations, resulting conflicts that undermine social systems, and the power arrangements that make exploitation possible.

Novel criteria are needed to develop a clearer understanding of what the dynamics and interaction between biobanking participants and researchers can mean in law. I put forward that such research relationships are continuous engagements that create obligations and expectations on both sides and should be recognized as such. So far, no comprehensive efforts have been made to develop precise criteria for addressing this although limited calls for the re-examination of gift relationships in genetic research have been made by law and philosophy scholars (Dickenson 2007, Gottlieb 2004, Knoppers 1998). They entertain preliminary reasons that lend support to a re-evaluation of human tissue gifts in law which include a desire for mutual relationships with obligations on both sides. In this light, a thorough rethink of the language of the gift in research is critical and long overdue. My analysis on the role of reciprocity is part of a broader proposal for the *empowerment* of group participants in research

Empowerment and the conditional gift

In seeking to establish reciprocity in the governance of group research, I put forward a new research ethics principle of group empowerment (Kanellopoulou 2008). *Empowerment* prescribes that participants are protected in mobilizing themselves efficiently and that their contribution to the research enterprise is acknowledged and respected. The principle places emphasis on the need to recognize the value of biobanking participants as partners and proactive contributors in research and to inscribe relevant research relationships in law as *continuous* and *reciprocal*. It focuses on the collaborative aspects of research relations and stipulates that there is a need for a balanced evaluation of the shared role of biobanking participants in the research process. The underlying necessity is the maintenance of an equitable balance between participants' and researchers' collaboration and the recognition of the continuous nature of their relationship as dynamic and worthy of protection. This approach respects the interface between participants and researchers as a rich site of sensitive social relations constitutive of a dynamic engagement. In offering ways to review and enhance participants' power and involvement in research, the concept of empowerment is inspired by notions of justice and fairness. These principles commend that the collective value of research participants is understood both as intrinsic as well as instrumental in the realization, facilitation and organization of long term research.

Empowerment proposes a new way to effectively regulate the interactions between biobanking participants and researchers. Biobanking research gifts should not be regarded as free gifts but instead as conditional gifts from participants to researchers, thereby affording the former some power in their ongoing involvement with research. I have envisaged a number of ways in which the empowerment principle can be integrated in legal mechanisms elsewhere as part of a detailed comparison of different possible mechanisms (Kanellopoulou 2008). For the purposes of this edited collection, I briefly discuss the conditional gift as a suitable regulatory option for participants' empowerment in research, facilitating contexts where they can no longer be viewed as vulnerable or invisible. Two possible interpretations of the term 'gift' can be distinguished and understood, only one of which would be best suited to address shifts in the research context. They both derive from an analysis of legal definitions but also from sociological and anthropological theory that regards gifts as necessarily involving relationships:

1. *Unconditional gift* can be defined as a gratuitous transfer with no expectation of return;
2. *Conditional gift* involves the exchange of reciprocal returns implied by ongoing collaborative interaction and dynamic relations between mutually engaged parties.

The notion of 'conditional gift' describes a different dynamic to the one manifest in unconditional interactions: unless balanced conditions that impose limitations on the uses of tissue in research are met by the parties, the gifts are not valid – in that case, cooperation and trust perish and the relationship does not exist. The notion of unconditional or free gift is better termed as 'donation' in law (Laurie 2002). It raises no obligation for reciprocation although sometimes it may still cause feelings of gratitude which nevertheless lie outside the scope of the law.

In assessing the appropriateness of conditions (for conditional gifts), new criteria could be developed that take account of: (1) the collaborative relationship between participants and researchers; (2) the nature of the gift which can include the significance of the transfer to participants' community; (3) participants' intentions for the use of the gift in particular ways; and (4) their expectations about possible benefits to be associated with the research, considered in the broader context of the reasons why a particular type of project is proposed. As a product of an ongoing dialogue, the conditional gift model is flexible and well-suited to protect biobanking participants' collective interests in how samples are used or what research is pursued with them. It can incorporate charitable intentions but also allows for mutual acceptance of conditions, together with collaborative assessment of biobanking participants' and researcher expectations. These conditions should include the recognition of participants' role and contribution in the biobanking process and respect to their values and diverse beliefs, where applicable. These can be understood broadly, in line with their history, connection with researchers, invested time, effort and motives. They should take account of participants' desires *for* and the limitations *on* the use of their samples, data and any other resources provided, including their possible contribution to funding and networking. This approach seeks to establish a more discerning understanding of research engagements in law and offers a flexible way for awarding biobanking participants with better levels of control in the research process.

Conceptually, the conditional gift is a reciprocal model for 'returning favours', a model of fair play (Schmidtz 2005). Returning favours means that participants and researchers do each other a favour by respecting, listening, understanding and trusting each other, by working towards common goals, cooperating, and, thus, empowering each other. Some philosophers consider 'duties of reciprocity' to be special as they arise from membership in a cooperative scheme. They contend that these duties should be distinguished from contracts because their very nature depends on returning the favour (Becker 1986: 73). Schmidtz (2005: 455) adds that duties of reciprocity should be distinguished from contractual obligations because they arise from implicit agreements. Arguably, this is where ideas about contracts as social metaphors become pertinent in the research context; they highlight parties' interests in clarifying the terms of research interactions and in negotiating conditions. I would not accept the use of reciprocity as a mere metaphor; reciprocity is conceptually stronger and more advantageous than the social contract metaphor in that it calls for shared and balanced returns in research engagements by reinforcing the collaborative qualities of the interaction – the conditions are to be assessed in proportion to the contribution of all parties involved in the research relationship.

Conclusion: On the Road to Biobanking Relationships

It is important to keep in mind previous references to reciprocity in human research ethics, however underdeveloped these may be. Reciprocity has been referred as 'a notion of exchange … [the] recognition of the participation and the contribution of the research participant' (Knoppers and Chadwick 2005: 75). These considerations are useful but do not provide guidance on how these elements are to be understood or evaluated. I have discussed that these need to be refined further by focusing on the character of their overall social relations. I have offered a way to establish reciprocity as a central value in regulating research under the principle of empowerment. Knoppers and Chadwick (2005: 76) understand that reciprocity recognizes autonomy, respects the personal and cultural values of participants and requires high levels of communication and transparency. These elements are also key in empowerment. I suggest that the principled use of reciprocity in biobanking should focus on the nature of the relationship between participants and researchers not as a mere exchange but as a collaborative engagement, sustained by social relations, personal and cultural values. Any possible criteria used to refine further the parameters of reciprocity should take account of the participants' role and contributions, their prior history, their expectations for the use of samples and their comparative expectations in research. Under these criteria, participants and researchers can agree to return conditional research gifts to each other in ways that meet continuing interests, do not harm participants, and do not impede research.

This work is part of a broader conceptual challenge to develop protections for groups in research. It calls for clear normative principles and new criteria to evaluate the role of biobanking participants as group participants in research. It proposes that the perpetuation of altruism as the only paradigm in the regulation of biobanking does not anticipate fundamental implications for participants' interests, becomes a boundary-setting exercise to avoid potential relevant claims, and relies on unchallenged assumptions of established trust.

This chapter proposes a new way for law to regulate the nature of the relationship between participants and researchers as collaborative, ongoing and reciprocal. It stipulates that increasing claims in research must be regulated in tandem with a broader understanding of the social, cultural, ethical and economic context within which biobanking takes place. It puts forward a novel approach in favour of empowerment and equity in balanced research relationships, against a background of the increasing economic value of human tissue and commercialization, by both public and private institutions. Under this new approach, research gifts should be considered as conditional and regulated in accordance with requirements of reciprocity. Overall, this work supports a more intuitive and interdisciplinary understanding of this sensitive and complex field.

Bibliography

Amani, B. and Coombe, R.J. 2005. The Human Genome Diversity Project: The politics of patents at the intersection of race, religion, and research ethics. *Law & Policy*, 27(1), 152-88.

Andrews, L. 2006. Who owns your body? A patient's perspective on *Washington University* v. *Catalona*. *J Law Med Ethics*, 34, 398-414.

Appadurai, A. 1986. Introduction: Commodities and the Politics of Value, in *The Social Life of Things: Commodities in Cultural Perspective*, edited by A. Appadurai. Cambridge: Cambridge University Press.

Bellivier, F. and Noiville, C. 2006. *Contrats et Vivant – Le Droit de la Circulation des Ressources Biologiques*. Paris: L.G.D.J.

Beyleveld, D. and Brownsword, R. 2004. My body, my parts, my property? *Health Care Analysis*, 8, 87-99.

Birks, P. 2000. The content of fiduciary obligation. *Israel Law Review*, 34, 3-38.

Blum, L. 1988. Review: [Untitled] Reciprocity. *Political Theory*, 16(1), 143-48.

Capps, B., Campbell, A.V. and ter Meulen, R. 2008. *Access to the UK Biobank Resource: Concepts of the Public Interest and the Public Good*, UK Biobank Ethics and Governance Council.

Carrier, J. 1991. Gifts, commodities, and social relations: A Maussian view of exchange. *Sociological Forum*, 6(1), 119-36.

Chalmers, D. and Nicol, D. 2004. Commercialisation of biotechnology: Public trust and research. *International Journal of Biotechnology*, 6 (2/3), 116-33.

Council of Europe Convention for the Protection of Human Rights and Dignity of the Human Being with regard to the Application of Biology and Medicine. 1997 (CETS No. 164).

Department of Health (UK) White Paper. 2003. *Our Inheritance, Our Future – Realising the Potential of Genetics in the NHS*. [Online] Available at: http://www.dh.gov.uk/en/Publicationsandstatistics/Publications/Publications

PolicyAndGuidance/DH_4006538?IdcService=GET_FILE&dID=7531&Rendition=Web.

Dickenson, D.L. 2004. Consent, commodification and benefit sharing. *Developing World Bioethics*, 4(2), 109-24.

Dickenson, D.L. 2007. *Property in the Body: Feminist Perspectives*. Cambridge: Cambridge University Press.

Douglas, M. 1990. Foreword, in *The Gift: The Form and Reason for Exchange in Archaic Societies*, edited by M. Mauss, London: Routledge.

Explanatory Report, Council of Europe Convention for the Protection of Human Rights and Dignity of the Human Being with regard to the Application of Biology and Medicine. 1997 (CETS No. 164).

Ferguson, J. 1988. Cultural exchange: New developments in the anthropology of commodities. *Cultural Anthropology*, 3(4), 488-513.

Frow, J. 1997. *Time and Commodity Culture: Essays in Cultural Theory and Postmodernity*. Oxford: Oxford University Press.

Gitter, D.M. 2004. Ownership of human tissue: A proposal for federal recognition of human research participants' property rights in their biological material. *Washington and Lee Law Review*, 61, 257-345.

Gottlieb, K. 1998. Human biological samples and the laws of property: The trust as a model for biological repositories, in *Stored Tissue Samples – Ethical, Legal, and Public Policy Implications*, edited by R.F. Weir. Iowa: University of Iowa Press.

Gouldner, A. 1960. The norm of reciprocity. *American Sociological Review*, 25(2), 161-78.

Greely, H.T. 2001. Human genomics research – New challenges for research ethics. *Perspectives in Biology and Medicine*, 44(2), 221-29.

Greely, H.T. 2005. DNA Fingerprinting and Civil Liberties Project Workshop Presentation, *American Society of Law, Medicine and Ethics*, Workshop 3.

Haddow, G. 2006. Because you're worth it? The taking and selling of transplantable organs. *Journal of Medical Ethics*, 32, 324-28.

Haddow, G., Cunningham-Burley, S., Bruce, A. and Parry, S. 2004. *Generation Scotland, Preliminary Consultation Exercise 2003-04, Public and Stakeholder Views from Focus Groups and Interviews*. Innogen Working Paper No. 20.

Haddow, G., Laurie, G., Cunningham-Burley, S. and Hunter, K. 2007. Tackling community concerns about commercialisation and genetic research: A modest interdisciplinary proposal. *Social Science & Medicine*, 64, 272-82.

Hardcastle, R. 2007. *Law and the Human Body: Property Rights, Ownership and Control*. Oxford: Hart Publishing.

Hardin, R. 1993. Altruism and Mutual Advantage. *Social Service Review*, 67(3), 358-373.

Hayashi, N., Ostrom, E., Walker, J. and Yamagishi, T. 1999. Reciprocity, trust and the sense of control: A cross-societal study. *Rationality and Society*, 11(1), 27-46.

Hayden, C. 2007. Taking as giving: Bioscience, exchange and the politics of benefit-sharing. *Social Studies of Science*, 37(5), 729-58.

Healy, K. 2006. *Last Best Gifts: Altruism and the Market for Human Blood and Organs*. Chicago: University of Chicago Press.

Healy, K. 2004. Altruism as an organizational problem: The case of organ procurement. *American Sociological Review*, 69, 387-404.

Human Genetics Commission. 2001. *Public Attitudes to Human Genetic Information – People's Panel Quantitative Study*. [Online] Available at: http://www.hgc.gov.uk/UploadDocs/DocPub/Document/morigeneticattitudes.pdf.

Human Genetics Commission. 2002. *Inside Information: Balancing Interests in the Use of Personal Data*. [Online] Available at: http://www.hgc.gov.uk/UploadDocs/DocPub/Document/insideinformation_summary.pdf.

Human Genome Organisation Ethics Committee. 2000. *Statement on Benefit Sharing*.

Human Genome Organisation Ethics Committee. 2007. *Statement on Pharmacogenomics (PGx): Solidarity, Equity and Governance*.

Hunter, K.G. 2006. DNA as taxable property – The elephant in the room or a red herring? *European Journal of Health Law*, 13(3), 263-75.

Kanellopoulou, N.K. 2004. Gift or duty? A normative discussion for participation in human genetic databases research, in *Blood and Data – Ethical, Legal and Social Aspects of Human Genetic Databases*, edited by G. Árnason, S. Nordal and V. Árnason. Reykjavik: University of Iceland Press.

Kanellopoulou, N.K. 2008. *Group Rights in Biolaw – A Model Approach*, PhD Thesis, University of Edinburgh.

Katz, R.A. 2006. The re-gift of life: Can charity law prevent for-profit firms from exploiting donated tissue and nonprofit tissue banks? *DePaul Law Review*, 55 (3), 943-1015.

Knoppers, B.M. 1998. Human genetic material: Commodity or gift? in *Stored Tissue Samples – Ethical, Legal, And Public Policy Implications*, edited by R.F. Weir. Iowa: University of Iowa Press, 226-35.

Knoppers, B.M. 2000. Population genetics and benefit sharing. *Community Genetics*, 3(4), 212-14.

Knoppers, B.M. and Chadwick, R. 2005. Human genetic research: Emerging trends in ethics. *Nature Reviews Genetics*, 6, 75-9.

Kollock, P. 1998. The anatomy of cooperation. *Annual Review of Sociology*, 24, 183-214.

Komter, A.E. 2005. *Social Solidarity and the Gift*. Cambridge: Cambridge University Press.

Kopytoff, I. 1986. The cultural biography of things: Commoditization as a process, in *The Social Life of Things: Commodities in Cultural Perspective*, edited by A. Appadurai. Cambridge: Cambridge University Press.

Laurie, G.T. 2002. *Genetic Privacy: A Challenge to Medico-legal Norms*. Cambridge: Cambridge University Press.

Levitt, M. and Weldon, S. 2005. A well placed trust? Public perceptions of the governance of DNA databases. *Critical Public Health*, 15(4), 311-21.

Lock, M. 1997. The alienation of body tissue and the politics of immortalised cell lines. *Body Society*, 7(2/3), 63-91.

Mahoney, J.D. 2000. The market for human tissue. *Virginia Law Review*, 86(2), 163-223.

Mason, J.K. and Laurie G.T. 2001. Consent of property? Dealing with the body and its parts in the shadow of Bristol and Alder Hey. *Modern Law Review*, 64, 710-29.

Mauss, M. 1990. *The Gift – The Form and Reason for Exchange in Archaic Societies*. London: Routledge.

McHale, J.V. 2004. Regulating genetic databases: Some legal and ethical issues. *Medical Law Review*, 12(1), 70-96.

Medical Research Council. 2000. *Public Perceptions of the Collection of Human Biological Samples*. [Online] Available at: http://ftp.cc.ic.ac.uk/pub/depts/neuropat/HS/humtis4a.pdf.

Medical Research Council Working Group on Human Tissue and Biological Samples. 2001. *Human Tissue and Biological Samples for Use in Research-Operational and Ethical guidelines*. [Online] Available at: http://www.mrc.ac.uk/Utilities/Documentrecord/index.htm?d=MRC002420.

Merz, J.F. et al. 2002. Protecting subjects' interests in genetics research. *American Journal of Human Genetics*, 70(4), 965-71.

Nicol, D. 2004. Public trust, intellectual property and human genetic databanks: The need to take benefit sharing seriously. *Journal of International Biotechnology Law*, 3(1), 89-103.

Nuffield Council on Bioethics. 1995. *Human Tissue: Ethical and Legal Issues*, London: NCB.

Oberman, M. 2006. When the truth is not enough: Tissue donation, altruism, and the market. *DePaul Law Review*, 55(3), 903-42.

Osteen, M. 2002. Introduction: Questions of the Gift, in *The Question of the Gift: Essays Across Disciplines*, edited by M. Osteen. London/New York: Routledge, 1-42.

Property Regulation in European Science, Ethics and Law Project (PropEur). 2006. *Final Project Conference Report*. PropEur Final Conference, Birmingham, 21-22 September 2006.

Pullman, D. and Latus, A. 2002. DNA Databases: A benefit-sharing model. *L' Observatoire de la Génétique*, 6.

Ratto, M. and Beaulieu, A. 2007. Banking on the Human Genome Project. *CRSA/RCSA*, 44(2), 175-200.

Schmidtz, D. 2005. What we deserve, and how we reciprocate. *The Journal of Ethics*, 9, 435-64.

Shaw, R. 2008. The notion of the gift in the donation of body tissues. *Sociological Research Online*, 13(8). [Online] Available at: http://www.socresonline.org.uk/13/6/4.html.

Sumner, J. 2007. *Public Attitudes to Biobanks and Related Ethics and Governance Issues*. [Online] Available at: www.egcukbiobank.org.uk/meetingsandreports.

Terry, S. et al. 2007. Advocacy groups as research organizations: The PXE International example. *Nature Reviews Genetics*, 8(2), 157-64.

Titmuss, R.M. 1970. *The Gift Relationship: From Human Blood to Social Policy*. London: George Allen & Unwin Ltd.

Tutton, R. 2004. Person, property and gift: Exploring languages of tissue donation. in *Genetic Databases: Socio-Ethical Issues in the Collection and Use of DNA*, edited by R. Tutton and O. Corrigan. London: Routledge.

Waldby, C. and Mitchell, R. 2006. *Tissue Economies – Blood, Organs and Cell Lines in Late Capitalism*. Durham, NC: Duke University Press.

Watkins, R.E. 2006. Negotiating rights and difference – Liberalism, cosmopolitanism and democracy. *Political Theory*, 34 (5), 628-33.

Winickoff, D.E. 2003. Governing population genomics: Law, bioethics and biopolitics in three case studies. *Jurimetrics*, 43, 187-228.

Winickoff, D.E. 2007. Partnership in UK Biobank: A third way for genomic property? *Journal of Law, Medicine and Ethics*, 35(3), 440-56.

Winickoff, D.E. 2008. *From Benefit Sharing to Power Sharing: Partnership Governance in Population Genomics Research*. JSP/CSLS Working Papers, Berkeley: University of California.

Chapter 3
From Benefit Sharing to Power Sharing: Partnership Governance in Population Genomics Research

David E. Winickoff

It is widely agreed that translating knowledge of the genome into clinical applications will require the construction of large searchable repositories of phenotypic information about patients – including disease diagnoses, environmental factors, and treatments – together with DNA sequences (Kohane and Altman 2005). These collections of DNA and biological information call for the sustained involvement of large numbers of research participants, and raise a host of difficult legal and ethical issues, especially when it comes to commercialization (National Bioethics Advisory Commission 1999, Greely 1999, Rothstein 2005). Whether or not participants in large-scale genomics projects are entitled, as an ethical matter, to enjoy some sort of direct benefits from the research – especially where commercial interests may be involved – remains a debated question (Berg 2001, Cambon-Thomsen 2004: 872, Andrews 2005, Hayden 2007). Benefit sharing proposals in population genomics have emerged in diverse contexts, from concerns about bioprospecting and the exploitation of indigenous groups, to the claims of disease group advocates, to debates over national genomic programs in Iceland, Estonia, Sweden, and the UK (Rothstein 2005: 95-7). For instance, in an early and influential statement on benefit sharing, the Human Genome Organization (HUGO) Ethics Committee draws upon commitments to the genome as common heritage of human kind, and to the non-exploitation of actual participants, to recommend returning health benefits to participants and allocating 1 to 3 per cent of profit streams to public health goods (HUGO Ethics Committee 2000). A different context animates Jon Merz and his group at the University of Pennsylvania. Drawing upon the well-known cases of PXE International (Terry 2003), the Canavan's disease group dispute in Florida (Andrews 2006), and the Health Sector Database case in Iceland (Greely 2000, Pàlsson and Rabinow 2001, Winickoff 2006a, Fortum 2008), Merz et al. (2002) declare that there 'has been a market failure with respect to the value added to the research enterprise by patient and subject groups'. Accordingly, they argue that 'ways should be found to recognize and reward their contributions'.

Historically, the commercial value of medical records, bioinformation, and gene patents emerged after the regulatory structures of bioethics came into being (Winickoff 2003: 188-90). For some bioethicists and policy-makers, benefit sharing becomes a necessary corrective for a system that previously ignored important questions of distributive justice (Simm 2005). But as a norm, benefit sharing has also been framed in relation to a second goal. As Canadian jurist Bartha Knoppers has put it, benefit sharing seeks to avoid both '*biopiracy* with no return to benefits to the families or community', and '*commodification of the person* through payment for access to DNA' (Knoppers 2000: 212-4). Benefit sharing, in other words, seeks to address the inequities of an appropriation of resource value from research participants, but without recourse to *quid pro quo* payment. Searching for an elusive middle way on the contested question of bodily property,[1] benefit sharing attempts to stitch a distributive norm at the seam of the market and gift economies.

This chapter will argue that while benefit sharing should be applauded insofar as it attempts to submit relations of biocapital to new claims of *distributive* justice, the project is likely to fail without greater attention to issues of *procedural* justice. In particular, we need to pay more attention to the constitution of distributive power over resources for genomic research. This position requires specification and justification through three interrelated arguments. First, benefit sharing as a discourse tends to settle political questions of distributive agency – the power to make distributive choices – in ways that exclude research participants from governing the resources they help to create. This is problematic from the standpoint of both political legitimacy and empirical findings on the expectations of research participants. Second, the notion of 'partnership governance' presents a productive avenue for achieving a normative shift from 'benefit sharing', a distributive value, to 'power sharing', a procedural one. If implemented, partnership governance would empower participants to exert a share in distributive decision-making in return for contributing to the economic and social capital of the project. But it would also assign research participants with correlative obligations to other capital partners such as funders and researchers, and duties to uphold the charitable missions of biobanks. Third, and finally, implementing partnership governance might be easier than it seems: legal architectures already exist within charitable trust law and corporate governance that could help achieve 'partnership governance', and address obvious challenges such as agency gaps and collective action problems.

Distributive Agency and Political Legitimacy

In a recent essay on benefit sharing, anthropologist Cori Hayden (2007: 729-58) has pointed out that by positing norms at the level of ethics rather than formal rights, benefit sharing produces its own kind of political relations among participants, researchers, markets, and variously imagined publics. One thing it does is effectively silence the claims of individual property rights in excised

1 For a useful, if opinionated, review of different takes on how to walk this middle path, see Bovenberg 2005, Harrison 2002 and Gitter 2004.

tissue and bioinformation that, if recognized, might perform a much more radical type of redistributive project. In a suggestive turn of phrase, Hayden (2007: 748) counsels that we could 'make benefit sharing a problem about broader questions of distributive agency, rather than one that is narrowly focused on skewed ledgers of stakeholder interest, badly or better recognized'. This focus on the distributive agency of research institutions is critical in thinking about benefit sharing, for it turns us back to questions of power.

Benefit sharing ought to concern us from the normative perspective of political legitimacy, which tends to posit the importance of accountability and representation within processes of collective resource gathering and distribution. A significant amount of discretionary power sits within the concept of benefit sharing. First, consider the question of what types of benefits are even imagined: compensation in money or health care? Access to research resources? Revenue streams from licensing and intellectual property? Trickle-down streams of the innovation process? Second, consider the question of who exactly should benefit: individual participants? The research group? The political collective? Indeed, the substance and meaning of benefit sharing will depend on who has the ultimate power to answer these questions.

Even within well-governed biobanking projects that explicitly seek to capture value for the 'public good', consider the amount of distributional choice involved. Which public is being imagined, and at what scale of organization – community, nation, or globe? Generating a return to the public is often an explicit goal, but multiple and conflicting visions of how to manage a genomic biobank for the public good coexist, raising important questions. Does public good consist in the generation of a research resource accessible as a commons to a broad set of users? Or does it consist in the ability to spur local economic growth in the form of new biotechnology companies? How will intellectual property be allocated? And when does a definition of public benefit entail giving information and medical benefits back to participants?

This point captures some of my own unease with current discussions and proposals for benefit sharing. The amount of distributive agency entailed in the management of biobank resources raises important questions of who should exert this agency, and how exactly 'benefit sharing' and 'public benefits' should be constructed. Who sets the rules and policies? Following Elinor Ostrom, the theorist of common resource management, these might be labeled 'constitutional choices' in the design of a shared resource, i.e., choices concerning the rules for who has control over policy-making (Ostrom 1990: 50-55). And it is here where I believe current thinking about the governance of genomic capital might be re-imagined.

It seems that even as courts, lawyers and ethicists decide that individual volunteers and donors cannot exert legal rights of property in their materials once excised for research,[2] it remains up to some ill-defined agency to help ensure that a rather vague principle of benefit sharing is upheld. Benefit sharing as a discourse settles political questions of distributive agency in ways that exclude research participants from sharing in the distributive power over the resources they help to create. Given this range of choices available for distributing 'public benefit', we might be concerned about legitimacy in the organization of distributive agency, and think seriously about how the donor collective will be represented in resource decision-making.

The Question of Who Wields Distributional Agency

One way to handle the large amount of distributive agency entailed in the management of biobanking resources would be to do it through transparent expert committees. Consider the so-called Newfoundland and Labrador Model of benefit sharing that has been endorsed by Gill Haddow, Graeme Laurie, Sarah Cunningham-Burley and Kathryn Hunter (2007) – whom I will call the Edinburgh group – as a way of dealing with community concerns about commercialization in genomics research. In this model, developed by Pullman and Latus (2003), an additional expert committee is established to solicit and negotiate collective benefits that will flow from deals between the genomic biobank and commercial companies. Pullman and Latus (2003) tell us that this committee would consist of people with appropriate expertise in genetics, medicine, pharmacology, business, law, health policy, and medical ethics, appointed by various governmental and university authorities. The Edinburgh Group praises the principles underlying this model: distributive justice, the communal nature of genetic information, and the promotion of health as a public good (Haddow et al. 2007: 279). They argue that ethical oversight bodies, such as the Ethics and Governance Council in UK Biobank,[3] would be 'ideally placed to recommend the adoption of benefit sharing arrangements' (Haddow et al. 2007: 280).

The governance model here is far superior to many models elsewhere, where institutional owners of biobank resources, whether corporate or non-profit, make resource allocations tacitly. However, a central problem remains: the Pullman and Latus approach praised by the Edinburgh Group seems so concerned with avoiding individual claims to benefit that it also deprives donors from having any

2 The canonical case here is *Moore v. Regents of the University of California* [1990] Cal. 793 P.2d 479, 493-97. See also *Greenberg v. Miami Children's Hospital* [2003] S.D. Fla. 264 F. Supp. 2d 1064, 1074-76; *Washington University v. Catalona* [2006] E.D. Mo. 437 F. Supp. 2d 985.

3 UK Biobank is a project funded by the UK Medical Research Council and the Wellcome Trust that plans to collect medical data and DNA samples from 500,000 research participants. For more on UK Biobank, see project website at: http://www.ukbiobank.ac.uk/ [accessed 7 October 2009].

real control, either as individuals or as a collective group, in the dispensation of the biovalue they have been so instrumental in creating. In other words, what these models seem to ignore is the ability of individuals and collectives of volunteers to have a say in what sorts of charitable uses should be preferred. In a field in which values surely conflict regarding the allocation of collective goods through biobanks, expert decision-making becomes less legitimate. These approaches fail to protect the control interest of charitable research participants in allocating access and use rights to their biomaterials and biological information. Because managing these resources entails choices between potentially conflicting goals, for example, between building local economic value and advancing global research access, this control interest becomes more important.

Recent empirical findings about attitudes of research participants toward commercialization underscore the importance of this control interest in bodily property and information. The Edinburgh Group found that research participants care deeply about maintaining their samples and information for public benefit, but that the issue of control is central to resolving their ambivalence about commercial access (Haddow et al. 2007: 277). In addition to identifying commodification as a concern, Haddow et al. (2007: 278) state that 'the root of public ambivalence [about commercialization] seems to lie in (i) notions of justice and fairness about private profit being made through public exploitation; and (ii) a perceived lack of control in terms of governance'. In these responses, the group discerns support for the control of biobank material and information by a public and trusted intermediary, and for the operation of the expert Ethics and Governance Committee within UK Biobank for handling allocation decisions in commercial relations. My own view, which I have expressed in writing elsewhere (Winickoff 2007: 446-49), is that an opportunity is missed for drawing a more radical conclusion from this research: namely, that the volunteer participants within such projects would desire and deserve some sort of role in distributive decision-making. Control is a major concern for research participants, so that addressing control – not just distribution – might be an important pragmatic goal for preserving trust and interest.

Thus, in my view, procedural justice and pragmatism both require the constitution of real rights not of benefit, but of partial control of biobanks as common-pool resources. While this approach has its own challenges and problems, I also believe this approach to benefit sharing is more legitimately generalized across situations and cultures, both in developed and developing countries.

Partnership Governance: Incorporating Research Participants

In the course of the well-documented failures of the Human Genome Diversity Project, community participation in research governance of population genetics projects emerged as a central concern. Bioethicists involved in the project started to articulate the importance of community consultations as a mechanism to protect the collective interests of the researched group (North American Regional

Community of the Human Genome Diversity Project 1997). The mechanisms for community consultation ultimately failed to satisfy many indigenous groups that had been approached for sampling, and the Human Genome Diversity Project eventually collapsed.[4] Community consultation has survived as an important norm in population genetic research projects (Foster et al. 1999, International HapMap Consortium 2004), but the notion still fails to solve important ethical and legal problems of providing legitimate group representation (Juengst 2000).

Drawing from my own comparative study of genomics, I have outlined new representational legal forms for the donor collective in biobanking that would carry both social and scientific benefits for research, and that hold promise for negotiating a better path between raw commodification and market inalienability of human tissue collections (Winickoff and Winickoff 2003: 1180-1184, Winickoff and Neumann 2005, Winickoff 2006b). Building on this previous work, and the selected work of others, I believe that a notion of 'partnership governance' could enhance the value, viability and legitimacy of genomic biobanks. Partnership governance seeks to go further than existing mechanisms of 'community consultation' by implementing control rights at the level of the research participant collective. Its mechanisms would be the aggregation of individual rights of control through private ordering or contract in order to construct a shared governance structure over a collective genomic resource.

Others have theorized some interesting ways to reconstitute the role of participants in biobanks. In a 2003 essay, anthropologist of science Mike Fortun invokes the notion of trade unionism, via Polish solidarity,[5] to suggest ways in which the politics of biobanking could be re-imagined from the ground up. There, Fortun (2003) argues that a bioethics structured to protect individual autonomy and privacy in biobanking has the unfortunate add-on effect of reproducing the 'atomization that benefits the status quo alone'. He asks, 'what are the means by which the participants in these efforts – the people who provide the informative flesh, without which nothing would be possible – will be given a *collective* voice in the future of the enterprise?' (Fortun 2003: 3).[6] Stressing the need for a more democratic biopolitical order, Fortun (2003: 3) argues that organized labor might be a useful model for producing collective education, deliberation, and decision-making.

4 For an important account of the rise and fall of the Human Genome Diversity Project, see Reardon 2004.

5 Solidarity was the name of the famous trade union movement that helped overthrow the ruling socialist state in Poland. (See, e.g., Ash 2002).

6 Fortun invokes solidarity in a different way than bioethicists Chadwick and Bere in the UK Biobank context. These bioethicists use solidarity to emphasize the communitarian obligations of individuals towards the collective public health and biomedical research enterprise (Chadwick and Bere 2001: 318-21). Whereas Chadwick and Bere use solidarity to make a communitarian defense of open-ended consent so that biobank research can proceed with less hindrance, Fortun uses it with Marxian and Habermasian inflections to assert the need for enhanced class-consciousness and democratic deliberation.

These are powerful goals, and trade unionism is an interesting paradigm for putting them into practice. But why should the analogy necessarily be labor instead of capital?[7] After all, the donors are providing the physical and information capital necessary for the collective resource to be built. It is true that the value of biobanks is enhanced if participants are willing to be recontacted, and to update their information over time, and that this could be construed as a form of labor. However, looking at the situation prior to donation and the transfer of entitlement, the group of donors as a collective possesses a crucial form of material, informational and biological capital that could be used to demand a share of power. This is one of the insights to be drawn from the PXE International story, where disease group members formed and retained legal control of a biobank in order to help advance the particular research goals of the organization (Winickoff 2003: 222-26).

I prefer the notion of 'partnership governance' as a promising principle of power sharing, because it would recognize the contributions of participants as capital, not labor, conferring upon the participant group the legal standing to share in distributive decision-making. The notion of 'partnership' works against the idea of exclusive ownership, and at its core connotes a form of cooperative human relations with respect to shared conditions and mutual aims. In modern vernacular usage, a partner is 'a person with a joint share in or use of something' (Simpson et al. 1989).[8] We speak of spouses and intimate relations as 'partners'. In the business law context, partners are 'individuals with interests and investments in a business or enterprise, among whom expenses, profits, and losses are shared' (Simpson et al. 1989).[9] In both the American and British legal systems, partnership involves joint control of assets, and an equity interest in risks and benefits.[10] A legal partnership also entails mutual fiduciary duties of loyalty. In one judge's famous words, what partners owe each other is 'not honesty alone, but the punctilio of an honor the most sensitive'.[11]

But negotiating the politics of genomic capital will require moving beyond rhetoric to construct a legal architecture for partnership governance. In previous work setting out what I have called the Charitable Trust Model, I have tried to show how this could operate through what are called donor-advised trusts (Winickoff and Winickoff 2003, Winickoff and Neumann 2005). I won't review those proposals here, but instead address important criticisms I have received from friends and colleagues about these ideas. While participatory governance might work for culturally and politically cohesive groups, the criticism goes, achieving it in larger

---

7 James Boyle makes this point in his useful discussion of John Moore's famous spleen (Boyle 1997: 97-107).

8 'Partner, n.(1)', first definition.

9 'Partner', fourth definition.

10 See, e.g., *Uniform Partnership Act* [1997] (US) § 18 (a) (e); and *Partnership Act* [1890] (Eng.) (2)-(7).

11 The words are Judge Benjamin Cardozo's in the case of *Meinhard v. Salmon* [1928] N.Y. 164 N.E. 545, a foundational case in the law of partnership in the United States.

projects with many heterogeneous participants would be virtually impossible: such groups are too large to be meaningfully represented, and their interests are too disparate. It is one thing to see how a group like PXE International or an American Indian Tribe with sovereign legal status might possess the necessary level of political cohesion of purpose to produce a biobanking collective that is able to develop mutually acceptable terms of access and control to bodies, DNA, and health information. It is another to see how these examples are either relevant or useful in thinking through mechanisms to close the agency gap in large-scale projects (Bovenberg 2005: 931-32). How then to move 'partnership' from rhetoric to practice in large projects involving a diverse group with varied interests and purposes?

While administering such partnership governance in a meaningful way would be a complex task, I propose that principles drawn from the law of corporations should help solve some of the most difficult collective action problems. This is less strange than it may seem, for the corporate form is applicable to organizing collective action not only for profit, but also for charitable goals. Could it be that the law of corporations offers us a way to construct 'the equity' of biocapital? I think so. Here, I refer to equity both in the sense of justice, and in the sense of capital ownership.

Each individual donor's stake in the management of a genomic resource is too small to warrant care, but the sum total is valuable and donors are likely to have preferences about its charitable distribution. This is a classic collective action problem, and a familiar one in the realm of corporate governance in the for-profit arena, where a single corporation might have millions of capital contributors, i.e., the shareholders. An individual shareholder might not have sufficient time or interest to attend to matters of corporate policy, but the idea that shareholders will be adequately represented in corporate decision-making is one of the pillars of corporate governance.

But what about the important issue of control, and the potential gap between the donor group and their representatives? Much of the law of corporations is devoted to solving problems of agency that can arise between managers of pooled assets, and the shareholders. Rules about fiduciary duties of managers, laws against self-dealing, and proxy voting are all aimed at closing the gap between managers and shareholders, a relationship that is fraught with the potential for mistrust and misappropriation. Why shouldn't we bring the same governance arsenal to bear in the realm of charitable biobanking? Hopefully, I have argued successfully that there is neither a theoretical nor a pragmatic reason against doing so – to the contrary, there is every reason to do so.

Such an approach would necessarily have to be adapted to local circumstances, but let's play out the idea in the context of UK Biobank, a project planned to include half a million people. UK Biobank is already set up as a non-profit corporation, UK Biobank Ltd., with a Board of Directors that makes major institutional decisions and sets project policy. UK Biobank managers might begin to implement partnership governance by taking a lesson from the formation of a trade union. During the consent process, potential donors would be informed that a Participants Association would be formed, and that they may sign on to the association as a voluntary matter if they chose to do so. Signing on would mean that they would be responsible for voting for Participant Association leadership, including a President. In order to help institute real power sharing within the biobank governance structure, this President would serve on the UK Biobank Board of Directors, akin to how a major institutional investor would sit on such a commercial corporate board. Furthermore, the Participant Association would be responsible for filling a number of seats on the Ethics and Governance Council and the Institutional Review Board, or perhaps form a separate Approval Committee. Once a certain threshold number of signatures were obtained, say 10,000 for the projected 500,000 person biobank, UK Biobank Ltd. would contact this group and notify them that the Association had been formed, and make a call for nominees for Participant Association leadership.

Each signatory to the Participant Association moving forward would have a voting share in the association elections. Voting could be facilitated by a combination of e-mail and regular mail, just as proxy voting occurs within corporate governance. UK Biobank would have to provide support in the form of a staff member who would engage exclusively as a participant liaison and relations officer. An interim leadership group would be appointed in this preliminary process, to be replaced or reconfirmed by annual elections in the future.

It would be the task of the Participant Association leadership to organize at least one public meeting per year, so that attitudes and preferences of donors could be assessed, and policy choices regarding resource distribution deliberated. For instance, this group could develop guiding criteria for preferable research topics in the biobank, and could also organize presentations from representatives from the other major funders to discuss joint goals and vision. The technical aspects of biobank decision-making pose significant challenges to generating meaningful discussions among a lay group of participants. However, models for deliberative public engagement on biobanks have already emerged (Burgess, O'Doherty and Secko 2008, Haddow 2008), and these would be a useful resource for designing productive and legitimate meetings, whose outcomes would help guide the Participant Association leadership.

Leadership would then be bound to represent these collective decisions on the Board of Directors and on the Ethics and Governance Council. Acting as both representatives of the donor group, and as officials of a charitable organization with a particular mission, their own charge would be to represent the donor collective as UK Biobank makes distributive and ethical decisions with respect to the use of the common pool resource, the biobank.

Addressing Obvious Objections

There are obvious objections to this proposal. First, there is the concern that a 'shareholding' discourse will bring volunteers out of the altruistic mode and put them into an interest group model in which they merely advance their self-interests. Wouldn't this talk of 'shareholders' and power sharing undermine the very altruism or sense of obligation that motivates people to participate in the first place?

It is true that the representatives of the Participants Association could help enact the collective preferences of the donor group with respect to how resources are allocated. However, this only underscores the importance of using the law of charitable organizations as a legal structure for the biobank. The major funders of UK Biobank, for one, decided to use this structure. The law requires charitable corporations and trusts to be managed in accordance with their charitable missions. Neither the donor representatives, nor the Board of Directors as a whole, could act in a way that would benefit themselves or their groups in a direct financial way without jeopardizing the organization's public mission and tax-exempt legal status. Instead, managers of the resource share legal obligations to manage in accordance with the purposes of the organization, which puts a check on self-dealing.

More importantly, this objection rests on the mistaken notion that the exertion of agency in how charitable contributions are allocated somehow vitiates altruism or is contrary to the spirit of public mindedness. The contrary hypothesis, namely that increasing the agency of charitable donors might actually encourage more investment in public-minded projects, actually seems more likely. Large charitable donors, such as Bill and Melinda Gates and others, are motivated to make large charitable contributions in part by the fact that they can direct donations to what they see as the most pressing charitable priorities. Research participant representation within biobank management would not imply that they could or should advance narrow self interests; it only means that they have a share in determining how a collective public resource is charitably allocated, a form of empowerment that might actually enhance the spirit of public giving.

The second major objection to this proposal is that there are significant efficiency costs involved in implementing partnership governance: inviting research participant representatives to the management table would be difficult to achieve organizationally, and would make consensus more difficult.

Partnership governance certainly embodies a commitment to procedural justice, and this value can at times conflict with the important goal of administrative efficiency. However, there actually may be efficiency gains from partnership governance, especially in the longer term. The commitment to procedural justice in the biobanking context is based in part on the contention that the solution to difficult and even unanticipated problems in the governance of this fast moving technological field will require a robust, flexible, and appropriate constitution of power. For the oversight of a complex common-pool resource like a genomic biobank, building an architecture *ex ante* for the legitimate representation of research participants before challenges are encountered, is likely to be a good investment.

In other words, focusing on a constitutional powers approach, rather than fixed *ex ante* benefit share, to address the issue of participant interests is a way to preserve efficient and smooth operation in face of an evolving social compact (Winickoff and Neumann 2005). The sudden emergence of personal genomics companies and also the falling price of whole genome scans exemplify this field's fast-moving nature. With the rise of personal genomics companies like Navigenics and 23andMe (Goetz 2007), it may be necessary to renegotiate the terms of information provision back to biobank participants in non-profit contexts. Furthermore, the falling costs of whole-genome scans and their potential integration into research models has complicated the analysis of consent, privacy and benefit sharing in biobanking governance (Caulfield et al. 2008). As biobank managers begin to weigh whether and how to allocate biobank resources to information return, and make new decisions on privacy policies in face of new technological capabilities, representatives of the donor group would already be sitting at the management table. Circumstances will change and the course will have to be altered. Partnership governance may provide an efficient mechanism to handle controversial choices as the managers of genomic capital have to change course in response to new scientific and economic conditions. A commitment to fair decision-making would enhance collective trust in the face of tough policy decisions, a crucial end in itself for research with human participants (O'Neill 2002). But this speaks to the efficiency objection as well. A behavioral economist might tell you that increasing the agency of charitable donors might actually encourage more investment in public-minded projects.

Conclusion

To date, the governance of genomic biobanking as well as human subject research has tended to ignore the collective interests of research participants, instead focusing on individual consent and privacy issues. Norms of 'benefit sharing' and 'community consultation' have emerged to help fill a policy lacuna that keeps widening as biomedical research becomes more and more integrated into commercial markets. But the concerns of 'distributive agency' articulated by Cori Hayden in her critique of benefit sharing become only more important in an era in which public-private collaborations have become the norm. Governance mechanisms in human subject research not only protect individuals from potential harms, but also tacitly order the power to make distributive decisions over valuable collections of informational, genetic, and social capital. While justice has been a traditional concern within bioethics, current conditions within the research environment call for its specification as not just a distributive, but also a procedural, norm.

What I am arguing for is a different constitution for biobanking, one that moves to the procedural axis of justice by building rules for 'power sharing' rather than mere 'benefit sharing'. Principles and legal forms derived from charitable trust law and corporate governance, such as trusteeship, fiduciary duties and shareholding, could be useful for solving problems of collective action, representation and agency involved in that shift. Partnership governance could establish appropriate power conditions for negotiated solutions to distributional dilemmas, without recourse to pure market solutions in which individuals are forced to negotiate for benefits against large rent-seeking institutions. Such an approach would be democracy enhancing, and might construct a useful architecture for the new understandings of biological citizenship[12] that are emerging along with genomic sciences. Without upsetting the public mission of research, power sharing through partnership governance would begin to flesh out process rights to biological citizenship and provide avenues to reorder the relations between individuals and their medical authorities.

Bibliography

Andrews, L.B. 2005. Harnessing the Benefits of Biobanks. *Journal of Law, Medicine and Ethics*, 33, 22-28.

Andrews, L. 2006. Who Owns Your Body? A Patient's Perspective on *Washington University* v. *Catalona*. *Journal of Law, Medicine and Ethics*, 34, 398-407.

Ash, T.G. 2002. *The Polish Revolution: Solidarity*. New Haven: Yale University Press.

Berg, K. 2001. The Ethics of Benefit Sharing. *Clinical Genetics*, 59, 240-243.

Bovenberg, J. 2005. Whose Tissue is it Anyway? *Nature Biotechnology*, 23, 929-33.

Boyle, J. 1997. *Shamans, Software and Spleens: Law and the Construction of the Information Society*. Cambridge: Harvard University Press.

Burgess, M. O'Doherty, K. and Secko, D. 2008. Biobanking in British Columbia: Discussions of the Future of Personalized Medicine through Deliberative Public Engagement. *Personalized Medicine*, 5, 285-296.

Cambon-Thomsen, A. 2004. The Social and Ethical Issues of Post-Genomic Human Biobanks. *Nature Reviews Genetics*, 5, 866-73.

Caulfield, T. et al. 2008. Research Ethics Recommendations for Whole-Genome Research: Consensus Statement. *PLoS Biology*, 6(3), e73.

Chadwick, R. and Bere, K. 2001. Solidarity and Equity: New Ethical Frameworks for Genetic Databases. *Nature Reviews Genetics*, 2, 318-21.

Fortun, M. 2003. Towards Genomic Solidarity: Lessons from Iceland and Estonia. *Open Democracy*. [Online] Available at: http://www.opendemocracy.net/content/articles/PDF/1344.pdf [accessed 1 October 2008].

12 For a review on the emerging literature on 'biological citizenship', see Rose 2008: 6.

Fortun, M. 2008. *Promising Genomics: Iceland and deCODE Genetics in a World of Speculation*. Berkeley: University of California Press.

Foster, M.W. et al. 1999. The Role of Community Review in Evaluating the Risks of Human Genetic Variation Research. *American Journal of Human Genetics*, 64, 1719-27.

Gitter, D.M. 2004. Ownership of Human Tissue: A Proposal for Federal Recognition of Human Research Participants' Property Rights in Their Biological Material. *Washington and Lee Law Review*, 61, 257-345.

Goetz, T. 2007. 23AndMe Will Decode Your DNA for $1,000. Welcome to the Age of Genomics. *Wired Magazine*. [Online], 15. Available at: http://www.wired.com/medtech/genetics/magazine/15-12/ff_genomics [accessed 1 October 2008].

Greely, H.T. 1999. Breaking the Stalemate: A Prospective Regulatory Framework for Unforeseen Research Uses Of Human Tissue Samples and Health Information. *Wake Forest Law Review*, 34, 737-66.

Greely, H.T. 2000. Iceland's Plan for Genomics Research: Facts and Implications. *Jurimetrics*, 40, 153-91.

Greenberg v. Miami Children's Hospital [2003] S.D. Fla. 264 F. Supp. 2d 1064.

Haddow, G. et al. 2007. Tackling Community Concerns About Commercialisation and Genetic Research: A Modest Interdisciplinary Proposal. *Social Science & Medicine*, 64, 272-82.

Haddow, G. et al. 2008. Generation Scotland: Consulting Publics and Specialists at an Early Stage in a Genetic Database's Development. *Critical Public Health*, 18, 139-49.

Harrison, C.H. 2002. Neither Moore Nor the Market: Alternative Models for Compensating Contributors of Human Tissue. *American Journal of Law and Medicine*, 28, 77-105.

Hayden, C. 2007. Taking as Giving: Bioscience, Exchange, and the Politics of Benefit sharing. *Social Studies of Science*, 37, 729-58.

HUGO Ethics Committee. 2000. Statement on Benefit Sharing. *Clinical Genetics*, 58, 364-366.

International HapMap Consortium. 2004. Integrating Ethics and Science in the International HapMap Project. *Nature Review Genetics*, 5, 467-475.

Juengst, E. 2000. What 'Community Review' Can and Cannot Do. *Journal of Law, Medicine and Ethics*, 28, 52-54.

Knoppers, B.M. 2000. Population Genetics and Benefit Sharing. *Community Genetics*, 3, 212-14.

Kohane, I.S. and Altman, R.B. 2005. Health-Information Altruists – A Potentially Critical Resource. *New England Journal of Medicine*, 253, 2074-77.

Meinhard v. Salmon [1928] N.Y. 164 N.E. 545.

Merz, J.F. et al. 2002. Protecting Subjects' Interests in Genetics Research. *American Journal of Human Genetics*, 70, 965-71.

Moore v. Regents of the University of California [1990] Cal. 793 P.2d 479.

National Bioethics Advisory Commission. 1999. *Research Involving Human Biological Materials: Ethical Issues and Policy Guidance.* Rockville, MD: National Bioethics Advisory Commission.

North American Regional Community of the Human Genome Diversity Project. 1997. Proposed Model Ethical Protocol for Collecting DNA Samples. *Houston Law Review*, 33, 1431-73.

O'Neill, O. 2002. *Autonomy and Trust in Bioethics*. Cambridge: Cambridge University Press.

Ostrom, E. 1990. *Governing the Commons: The Evolution of Institutions for Collective Action*. Cambrdge: Cambridge University Press.

Pálsson, G. and Rabinow, P. 2001. The Icelandic Genome Debate. *Trends in Biotechnology*, 19, 166-71.

Partnership Act [1890]. (Eng.) (2)-(7).

Pullman, D. and Latus, A. 2003. Reconciling Social Justice and Economic Opportunism: Regulating the Newfoundland Genome' in Bartha M. Knoppers (ed.) *Populations and Genetics: Legal and Socio-Ethical Perspectives*, 543-566.

Reardon, J. 2004. *Race to the Finish: Identity and Governance in an Age of Genomics*. Princeton: Princeton University Press.

Rose, N. 2008. *The Politics of Life Itself: Biomedicine, Power, and Subjectivity in the Twenty-first Century*. Princeton: Princeton University Press.

Rothstein, M.A. 2005. Expanding the Ethical Analysis of Biobanks. *Journal of Law, Medicine & Ethics*, 33, 89-101.

Simm, K. 2005. Benefit Sharing: An Inquiry Regarding the Meaning and Limits of the Concept in Human Genetic Research. *Genomics, Society and Policy*, 1, 29-40.

Simpson, J. et al. (eds) 1989. *Oxford English Dictionary*. 2nd Edition. Oxford: Oxford University Press.

Terry, S. 2003. Learning Genetics. *Health Affairs*, 22, 166-71.

Uniform Partnership Act [1997] (US) § 18 (a) (e).

Washington University v. Catalona [2006] E.D. Mo. 437 F. Supp. 2d 985.

Winickoff, D.E. 2003. Governing Population Genomics: Law, Bioethics, and Biopolitics in Three Case Studies. *Jurimetrics*, 43, 187-228.

Winickoff, D.E. 2006a. Genome and Nation: Iceland's Health Sector Database and its Legacy. *Innovations*, 1, 80-105.

Winickoff, D.E. 2006b. Health-Information Altruists. *New England Journal of Medicine*, 354, 530-31.

Winickoff, D.E. 2007. Partnership in U.K. Biobank: A Third Way for Genomic Property? *Journal of Law Medicine & Ethics*, 35, 440-56.

Winickoff, D.E. and Neumann, L. 2005. Towards a Social Contract for Genomics: Property and the Public in the 'Biotrust' Model. *Genomics, Society and Policy*, 1, 8-21.

Winickoff, D.E. and Winickoff, R.N. 2003. The Charitable Trust as a Model for Genomic Biobanks. *New England Journal of Medicine*, 349, 1180-84.

PART 2
Consent

Chapter 4
Co-determination of Donors in Biobanks

Lukas Gundermann and Ulrich Stockter

Of all the legal and ethical aspects in biobanking, issues around consent are probably the ones that are discussed most intensely (Gibbons and Kaye 2007: 205). While there is a broad agreement about the need to obtain some kind of consent,[1] the type of consent required is hotly disputed (Elger and Mauron 2003: 270-275, Nõmper 2005: 45-80). Informed consent,[2] explicit consent,[3] open consent (Nõmper 2005: 80-144), broad consent (Kaye 2004)[4] and one-time general consent (Wendler 2006) are the variants in discussion. We will try to illustrate in this chapter that in our opinion, what is more important than the determination of a certain type of consent, is the adoption of a more holistic view that takes into account all relevant elements that safeguard the donor's interests in the biobanking context. While consent is certainly the cornerstone, other elements that add to the traditional concept of consent can help to achieve both the protection of the interests of the individuals involved and the proper functioning of a biobank. Our argument can be divided into two elements: Firstly, at a minimum level the donors' broad consent has to be obtained before their tissue and related data[5] can be processed. Secondly, wherever possible, the donors should be granted the possibility of exercising what we will call 'co-determination' as a means of safeguarding the donors' autonomy.

---

1 This is at least true with regard to Europe and the US. For a different approach in Western Australia see Laurie 2008: 186.

2 The classic standard in medical research, see Art. 22 of the Declaration of Helsinki of the World Medical Association; still supported by Caulfield 2007: 209.

3 Explicit consent is one possible ground for processing sensitive data according to Art. 8, para 2 a) of Directive 95/46/EC of the European Parliament and of the Council of 24 October 1995 on the protection of individuals with regard to the processing of personal data and on the free movement of such data (Directive 95/46/EC), the general date protection Directive.

4 Nõmper 2005: 76 labels this approach as 'broad consent with opt-out'.

5 As tissue can be regarded as a source of personal data, we will not distinguish between tissue and data in this chapter.

Consent as the Expression of the Donor's Autonomy

Autonomy and consent in Data Protection law

Autonomy is a crucial ethical principle in both medical law and data protection law (Laurie 2008). Autonomy grants individuals a legal status such that an act conducted in the absence of consent constitutes an unlawful infringement of their rights: *volenti non fit injuria*. In data protection law, consent is the first exception from the general prohibition on processing sensitive data.[6] In German law, the principle of autonomy in data protection law is backed by the constitutional right to informational self-determination. The entitlement to informed consent can also be derived from the principle of human dignity that is enshrined in the legal instruments on fundamental rights such as the Universal Declaration of Human Rights or the German constitution (Caulfield and Brownsword 2005: 73, Nõmper 2005: 117-118).

Part and parcel of the right to informational self determination is the requirement that the individual must be informed what data processing operations are to be performed using his or her data. Only if such transparency is achieved, is the individual in a position to determine the data use in question. Another important element of data protection law that can be related to informational self-determination is the so called 'purpose-binding principle'.[7] It requires the data controller to only process the data for the specific purposes for which they were collected. The purpose of the collection must be fixed in advance of the data collection. When collecting personal data from the data subject, he or she must be informed about the purposes and future recipients of the data.[8]

These legal standards obviously conflict with practical requirements of biobanks – and namely of population biobanks (Kaye 2004: 120). Since they are designed as resources for many future research projects, at the time of the donation no exhaustive description of the purpose of the future processing can be given. Accordingly, fully informed consent which generally demands information about purpose, risks and benefits of the treatment and the probability of their realization, cannot be obtained as required in accepted medical research practice. European data protection law, however, provides instruments for overcoming this hurdle (Kaye 2004: 123). The detailed information requirements may be restricted according to Art. 13 of Directive 95/46/EC under certain preconditions. Of most importance with regard to the consent requirement is Art. 8, para 4 of the Directive that entitles Member States to stipulate legal grounds for processing in addition to those laid down in the Directive. Such additional legal grounds for processing

6 See Art. 8, para 2 a) of Directive 95/46/EC.

7 Art. 6, para 1 b) of Directive 95/46/EC; second principle according to the Data Protection Act 1998, Schedule 1.

8 See Art. 10 of Directive 95/46/EC. The obligation does not necessarily comprise information about risks and benefits of using the data. See Kaye 2004: 119.

must meet certain preconditions: the respective processing operations have to be necessary with a view to a 'substantial public interest' and 'suitable safeguards' must be provided.

Thus, the general need for explicit informed consent can be abandoned if a substantial public interest so requires. According to the law, autonomy does not trump in every case. If we accept the view that biobanks are a scientific instrument that contributes to medical research in a substantial way and thus effectively serves the public good, the law permits autonomy to be overridden in favour of other ethical principles, namely beneficence.[9] It is important to point out at this stage, that the relevant provision of the Directive does not necessarily demand a black-and-white approach. It also justifies a reduced standard of informed consent, for example, allowing for a one-time broad consent plus additional elements that safeguard autonomy.

Is the possibility of opt-out sufficient?

Despite the legal possibility of using Art. 8 of the Directive, which would allow personal medical information to be used without consent if the right conditions are satisfied, there is broad agreement that at least some sort of consent should always be obtained.[10] Some, however, argue that this minimal consent requirement does not have to be a positive consent but could just be the possibility to 'opt out' of comprehensive biobanking research schemes (Elger and Mauron 2003: 282-90, Johnsson et al. 2008).[11]

This position appears to have been supported by an empirical study conducted in Sweden which found that informed consent requirements for biobank research are of relatively low interest to the donors and the vast majority of donors would agree to unspecified future uses of their data for medical research purposes (Johnsson et al. 2008: 225-26, Wendler 2006: 546, contra: Caulfield 2007: 218-222, who provides a more nuanced account of public perception of genetic research). Johnsson et al. (2008) argues that consent requirements place a heavy burden on biobanks, since they demand the establishment of a costly management system. Given the public good that is being pursued by maintaining a biobank, it would be disproportionate to impose such high costs as only very few individuals make use of it (Wendler 2006: 546) so that they might even amount to a legal formalism.[12] Thus, an active opt-in system would be unnecessary; it would be sufficient to provide an opt-

9 In this context, Laurie 2008 points to the legal situation in Western Australia, where there is not even the possibility of opting out of the use of one's medical data for research purposes.

10 See above.

11 An opt-out scheme was also the basis of the comprehensive Icelandic health sector database, see Gertz 2004.

12 'Bureaucracy with no benefits' (Johnsson et al. 2008: 226).

out possibility (Johnsson et al. 2008: 226). Such proposals, however, have been received with some reservation (Laurie 2008).

The argument brought forward by Johnsson et al. (2008) is based on the conclusion that the granting of a certain right might not in fact be necessary if a relatively small number of individuals actually exercise that right.[13] This approach seems to be based on a utilitarian view: In particular if the granting of such a right creates impediments for other legal subjects, the net benefit seems to be greater if that right is withheld. The result would obviously be different if, say, more than 50 per cent of the probands exercised their right (or at least believed such a right was important). In our opinion, however, the restriction or withholding of a right must not be based on empirical assumptions. The right to give or withhold consent opens up a margin of freedom for the individual: they may exercise the right or abstain from doing so. This legal position derives from civil liberties (or fundamental rights) which determine that the state is not permitted to interfere unless there is a special justification. Accordingly such a right may be restricted when being reconciled with other legal positions of the same nature (for example freedom of research). However, the argument that the right is exercised only rarely is not sufficient justification for restricting it. In other words: one may not be deprived of the right to demonstrate just because so far one has never taken to the streets.

Quality of genetic data

A lower standard of consent with regard to biobank research is sometimes considered sufficient based on the notion that such research only entails minimal risks[14] for the individual (Robertson 1999: 67-68). Accordingly, the abandonment of consent as a requirement is justified. However, for this to hold true there must be little possibility that the material in biobanks would be misused and there would have to be 'strong laws against unauthorized disclosure' (Robertson 1999: 68). Even though we do not wish to promote genetic exceptionalism, genetic data are unlike other types of data as they provide a range of very sensitive information and it can be argued that they continue to hold the status of personal data, even after anonymization (Kaye 2004: 127-128). In the case of biobanks, where new personal information is added to DNA data (or tissue that can be used to generate such data) over a long period of time, the risk of being able to identify an individual will increase. We would argue that such risks are comparable with the ones resulting from (at least smaller) medical procedures. In the long term, the knowledge of an individual's genetic make-up may lead to greater harm if it is misused than is the case with a temporary physical condition resulting from a medical intervention. There may also be monetary issues at stake in the use of human tissue, as the

13 See also Caulfield 2007: 222-224.

14 For an overview of the presumed 'minimal risk' as a justification for research without consent, see Nõmper 2005: 58-61.

case of *Moore v. Regents of the University of California* (51 Cal 3d 120 (1990)) illustrated. For good reasons informed consent is accepted unchallenged as a legal precondition for any bodily intervention in medicine and medical research be it ever so small. Considering the comparable risks associated with the processing of genetic data, we would contend that consent as a general precondition is of the same crucial importance in this area.

Therefore, some type of basic consent should always be obtained – an opt-out mechanism would not be sufficient. Before giving his or her consent, the donor should be informed about the general risks that may result from his or her donation – even though not every aspect of the future processing of his or her data can be anticipated. If he or she consents in principle to donating despite some risks, as an autonomous agent he or she must be taken seriously, thus satisfying an ethical demand. From a legal perspective, it will then depend on the additional safeguards that have been implemented whether the requirements laid down in Art. 8, para 4 of Directive 95/46/EC are met thus justifying a reduction in the standard from fully informed to basic consent.

Other arguments in favour of consent

In our view, the basic consent requirement serves additional functions apart from safeguarding autonomy. It can be used as a means of self-assurance and self-reflection for the biobank operators and researchers. In this regard, the provision of the relevant information to the donor in the context of obtaining consent can also be perceived as an important opportunity for the biobank or the researcher to receive feedback at an early stage and thus as a chance to check the assumptions and rationales that underpin the research protocol. In an ideal case, when approached for their consent, potential donors respond by raising questions or concerns relating to the information provided to them. Ideally, this could even amount to a process of shared decision making. In addition, other entities that are involved, such as research ethic committees, provide an important role in assessing research proposals.

The process of obtaining consent can also be seen as a means for generating mutual trust and confidence between donor and biobank.[15] While on an ethical level, autonomy and dignity underline the need for consent, this translates on a psychological and more pragmatic level into the need to address the donor not as an indifferent supplier of research tissues but as a partner, who by his or her donation enables research. In the long run, the communication between members of society and the biobank in the context of obtaining basic consent may also have positive consequences for the general acceptance of biobank research in society, continued political support and commitment to funding.

---

[15] For the importance of public confidence and support of biobanks, see Gibbons and Kaye 2007: 203.

It is not a new phenomenon that the adherence to certain standards or norms that appear to serve third parties in the first place will also have positive effects for the person or organization applying the norm. An example would be the insistence on medical confidentiality enshrined in the Hippocratic Oath. This principle was not developed as an act of pure unselfishness by the ancient physicians in the first place. In *On Decorum*, Hippocrates explicitly points out, that the physician who breaches the duty of confidentiality not only has a detrimental effect on himself but also the whole community of physicians. By the same token, the trust generated by obtaining a declaration of basic consent after duly informing potential participants about the details of the research that are known at the time is likely to lead to an increase in donations. On the other hand, as in the case of the ancient physicians, single incidents of a biobank's or researcher's failure may discredit the whole research community.

Result: (Basic) consent indispensable

As a consequence, the general requirement of consent for data and tissue usage in biobanks should be upheld (Kaye 2004: 130). Donors should be informed as precisely as possible about future uses and approached for the respective consent. If it has been possible to obtain real informed consent to well defined and detailed uses of data and tissue, such consent also serves as a boundary for future usage. Even though it might result in considerable efforts to approach the donors anew, fresh consent must be obtained if later, new research is proposed that differs from that to which the donor originally consented.

Moreover, in such a scenario, the informed consent obtained in the first place should be interpreted as conclusive. By obtaining precise informed consent the biobank implicitly waives the right to base further processing on any (possibly existing) statutory provisions which might justify such processing without consent. To do so, would amount to a deliberate deception of the donors, thus acting in bad faith. Good practice would require that donors should be informed that further processing would be carried out on the basis of the statutory provisions, which would enable them to make an informed decision about participation in the biobank.

This would apply particularly to population biobanks where the strict requirements of prior informed consent cannot be met. In order to safeguard the donor‘s autonomy, a two-stage approach should be followed. At the first stage the donor's general broad consent to the use of his or her data and tissue in the biobank for future research should be obtained (Kaye 2004: 131). The second stage compensates for the non-existence of a fully detailed informed consent by offering the donor the ability to decide, to a certain extent, on the future use of his or her tissue samples and data. We refer to this approach as 'co-determination'.

Donors' Co-determination

This term originally refers to the field of (German) labour law[16] and denotes the transfer of democratic decision making processes into private or public management structures. In this chapter, the term co-determination is used in a broader sense. It also includes less formalized ways of participation in decision making processes by individuals who are outsiders to the respective organization and may not even be identified as stakeholders in the genuine sense. Such participation is less formal than in the case of co-determination in labour law. Whilst it does not necessarily require group organization, it offers a way to integrate unorganized and changing streams of interest into management processes. Co-determination in this sense can be found in various fields with more or less institutionalized structures. In this sense, too, market demand and supply can be understood as an instrument for exercising co-determination. By taking an informed decision about purchasing certain goods, the consumer is able to make decisions not only about his or her own consumption but also, to a certain extent, about what can be procured in the market. Examples can be seen in fair trade coffee, green power electricity-mix, non-child-labour-products etc.

In the context of biobanking, co-determination is designed to compensate for a seemingly inevitable degree of restriction of the donors' autonomy. Similar to the way demand driven co-determination is exercised by consumers in the market, here co-determination would entail the possibility of donors exerting influence on the precise use of their data and tissue in a biobank and particularly in a population biobank. Such a possibility would not alter the broad scope of the initial consent, as on the basis of this consent, data and tissue could be used for any type of future research that is unspecified at the time the consent was obtained. Thus, co-determination is not principally reducing the openness (and therefore vagueness) of the general consent. The implementation of co-determination does however require transparency for donors: they must be kept updated on current and envisaged future use of the material (Kaye 2004: 131). This could be achieved by giving information to the general public[17] or by informing the donors individually. Given the ready availability of electronic means of communication, neither would require disproportionate resources.

Based on such information, the donors should be offered different options to actually decide how their data is used. If the donors are perceived as partners with biobanks and recognized for their role in enabling research to happen, it is only logical to expand co-determination to more substantial aspects. One such aspect could be the option of temporary[18] or conditional limitations on consent. This

16 The German term is *Mitbestimmung*.

17 A good example of this public information policy is provided by the US National Institutes of Health. Their web site, www.ClinicalTrials.gov, supplies an overview of ongoing clinical trials.

18 See Kaye 2004: 131, who suggests the renewal of consent every five years.

would mean that donors would decide the substantial conditions that applied to the use of their data or samples. For example, a donor could stipulate that research could only be conducted if 10 per cent of biobank research is done for orphan diseases or if there is a concept of benefit sharing. The permitted research uses of data and tissue could also be limited. An example would be a prohibition on the use of material to conduct research for biological weaponry. Moreover, the transfer of material to certain countries could be prohibited or made dependent on the existence of certain standards of data protection in the importing state.

Other restrictions could be imposed regarding the procedural conditions that applied to the research process. For example, research using the respective donor's tissue would only be allowed following a positive assessment by a competent ethics committee or by research institutions which follow certain binding corporate rules. The adherence to such obligations could be enforced with penalty clauses. For practical reasons, such conditions and restrictions would have to be standardized and limited in number. With regard to population biobanks, this could be done by responsible bodies, such as research ethics committees, members of the public, representatives of patients, and doctors.

Co-determination in the above sense would not affect rights of withdrawal (Kaye 2004: 131). On the contrary, the donor should be allowed to exercise the right of withdrawal and require the deletion of data and the disposal of tissue samples at any time, without the need to give a specific reason. This is important as situations may arise that have the potential to have a significant, negative impact on the donor's privacy rights. Examples of such situations are access by third parties such as insurance companies; and the breach of contractual duties or security by the biobank. But allowing for the possibility of withdrawal is crucial apart from these specific circumstances: for it is only where the ability to re-assess and perhaps amend the original decision to donate to a specific biobank exists, that a market-like situation is created where donors, like consumers, can adjust decisions according to new information or different evaluations.

The concept of co-determination would require that respective samples and data of one individual would have to be identifiable in order for the donor to exercise their rights. At the same time, good biobanking governance requires the samples and data to be stored in an encrypted form so that they cannot be directly linked to the donor's name and other data that relate to his or her person. A trusted third party who controls the technical means to link from the research data to a natural person and vice versa would have to be involved. It should be pointed out that systems that can implement those requirements already exist and are available in the market (Reischl et al. 2006). They can also be used to establish a path of communication between the biobank and the donors, thus enabling the implementation of feedback mechanisms. Such instruments may as well be used in order to implement a co-determination scheme. At least with regard to population biobanks subsidised by public funding, such efforts appear to be indispensable. Firstly, public bodies have a special obligation to respect and implement positions deriving from fundamental rights;[19] such obligations should be handed down to institutions doing research with public money. Secondly, given that the political will continues to encourage biobank research, subsidies should be targeted at promoting the type of research that it is likely to attract more donors.

Concluding Remarks

The ideas promoted in this chapter adopt a broader approach in tackling the discussion about the appropriate type of consent for biobanking research. Given the current structure of biobanks, a relatively broad form of donors' consent in the initial phase appears inescapable. This can be compensated for, however, by co-determination. Donors are able to decide on acceptable usages of their tissue and data based on their understanding of the initial concept for the biobank. As different research areas emerge, the biobank will keep the donors informed about these new developments. On this basis, the donors may re-consider their decision, demand the exclusion of their samples and data from certain research projects or eventually withdraw their consent and have their tissue samples destroyed. Admittedly, this would require additional efforts on the part of the biobanks, but, perhaps modern data handling and communications technologies will greatly assist in this regard. What is potentially of equal importance, is that biobanks may themselves profit from such a system. The self-control mechanisms effected by donors' co-determination may facilitate a learning process that serves to develop and strengthen biobanking governance structures. Co-determination would not only create a new type of trusted relationship between donor and biobank, it would also increase the public acceptance of biobank research as a something that serves the public good. It is this public acceptance that is so important in the context of the socially sensitive nature of the research projects with which biobanks are generally concerned.

Bibliography

Caulfield, T. 2007. Biobanks and Blanket Consent: The Proper Place of the Public Good and Public Perception Rationales. *King's Law Journal*, 18, 209-226.

Caulfield, T. and Brownsword, R. 2005. Human Dignity: A Guide to Policy Making in the Biotechnology Era? *Nature Reviews Genetics*, 7, 72-76.

Elger, B. and Mauron, A. 2003. A Presumed-Consent Model for Regulating Informed Consent of Genetic Research Involving DNA Banking, in *Population*

19 See Section 6 of the Human Rights Act 1998, Art. 1 (3) of the German constitution (available in English at: http://www.bundestag.de/interakt/infomat/fremdsprachiges_material/downloads/ggEn_download.pdf).

and Genetics: Legal and Socio-Ethical Perspectives, edited by B. Knoppers. Leiden/Boston: Martinus Nijhoff Publishers, 269-295.

Gertz, R. 2004. *An Analysis of the Icelandic Supreme Court Judgement on the Health Sector Database Act*. SCRIPT-ed 1(2), 241-258. [Online] Available at: http://www.law.ed.ac.uk/ahrc/script-ed/issue2/iceland.pdf [accessed 16 March 2009].

Gibbons, S. and Kaye, J. 2007. Governing Genetic Databases: Collection, Storage and Use. *King's Law Journal*, 18, 201-208.

Johnsson, L. et al. 2008. Patients' Refusal to Consent to Storage and Use of Samples in Swedish Biobanks: Cross Sectional Study. *British Medical Journal*, 337, a345.

Kaye, J. 2004. Abandoning Informed Consent, in *Genetic Databases: Socio-ethical Issues in the Collection and Use of DNA*, edited by R. Tutton and O. Corrigan. London/New York: Routledge, 117-138.

Laurie, G. 2008. Consent for Biobanking: Lack of Dissent When Opting in Doesn't Necessarily Support 'Opt Out'. *British Medical Journal*, 337, 186-187.

Nõmper, A. 2005. *Open Consent – A New Form of Informed Consent for Population Genetic Databases, Tallinn*. [Online] Available at: http://dspace.utlib.ee/dspace/bitstream/10062/818/5/nomper.pdf [accessed 16 March 2009].

Reischl J. et al. 2006. Pharmacogenetic Research and Data Protection – Challenges and Solutions. *Pharmacogenomics Journal*, 6, 1-9.

Robertson, J. 1999. Privacy Issues in Second Stage Genomics. *Jurimetrics*, 40, 59-76.

Wendler, D. 2006. One-time General Consent for Research on Biological Samples. *British Medical Journal*, 332, 544-547.

Chapter 5
Developing an Appropriate Consent Model for Biobanks: In Defence of 'Broad' Consent

Margaret Otlowski

In addition to the numerous tissue and other DNA collections which exist in many hospitals, tissue and blood banks, pathology laboratories and the like, new and major prospective 'biobank' collections are being established. These are structured resources comprising collections of human genetic samples and relevant personal information (including health, family history, lifestyle etc.) (OECD 2008). The object of such collections is to create a resource for large scale, longitudinal genetic research projects to facilitate population health research through a better understanding of the role of genetics in disease. Over the last decade, a number of such population databanks or 'biobanks' have been established in several countries (Cambon-Thomsen 2004), including the Icelandic Health Sector Database, the Estonian Genome Project, UK Biobank, and the CARTaGENE project in Quebec, and others are planned including the Joondalup Family Health Study in Western Australia. Through the study of gene-environment interactions in populations over many years, such initiatives hold great potential for human genetic research and resulting advancement in the health of the population. Undoubtedly, however, they also pose a number of legal as well as ethical and regulatory challenges (Cambon-Thomsen et al. 2007). One such challenge relates to participant consent to biobanking which is the focus of this chapter.

There is growing realization that, from a practical point of view, obtaining specific consent from each participant for each separate research project for which their samples and information are to be used is not feasible in the context of large-scale, long-term biobanks which are created as infrastructure platforms for an indefinite number of future research projects. As a result, there is growing acceptance of 'broad consent', albeit at times reluctantly, as this may possibly be at the expense of important principles. This move towards broad consent is most commonly justified on the basis of the 'public good' (on the ground that too strict requirements for consent might hinder progress) and/or public opinion data (on the basis that the public supports broad consent) (Caulfield 2007, Caulfield et al. 2007). In this chapter, I want to make the case in defence of broad consent; instead of thinking of it as a less than satisfactory compromise, I want to argue that it can be an effective consent model in this setting. I seek to make this argument primarily on the grounds of principle, rather than expediency, although there is no denying that broad consent is a cheaper, more practical option for biobanks than a requirement of specific consent for each project. The crux of my argument is that the lack of specificity of consent which is inherent in the biobanking context where future projects are not determined at the time samples and information are collected, does not detract from the core principles consent aims to protect.

This chapter reviews the operation of consent principles as they apply to biobanking, with a view to arguing for a more liberal approach which allows for 'broad' consent to future use of samples. In doing so, my objectives are twofold: Firstly, to ensure that unnecessary impediments to research are removed, thus facilitating research participation, and secondly, to ensure respect for autonomous choices by research subjects. It is clear from these objectives that there are both public and private interests at stake: the public interest, and indeed public health dimension, in ensuring that genetic research can progress without undue hindrance, as well as the private interest in respect for subjects. However, these interests should not be seen as polarised or exclusive of one another because, importantly, there is also a public interest in ensuring that the rights of subjects are appropriately protected as their trust and participation is key to the successful research endeavour (Otlowski 2008). As Campbell contends (Campbell 2007), a central ethical feature of biobanks is their fundamental dependence on altruism and trust from those who donate samples and information, and this generates corresponding ethical obligations on biobank creators and custodians to honour their commitments to donors.

Overview of Consent Principles

There are well established principles with regard to consent derived from the *Nuremberg Code* (1947) and *Declaration of Helsinki* (1964) which have been described as the 'gold standard' of research ethics (Kegley 2004). Essentially, they require the provision of information and the capacity of individuals to make a voluntary choice. Informed consent is seen as the basis upon which an ethical relationship between a participant and researchers can be negotiated; it plays an important practical, but also symbolic role in protecting and respecting the individual participant (Chadwick and Berg 2001). In the research setting, consent would be expected to cover *inter alia* research aims, methods, demands, risks and harms, inconveniences, discomforts, possible outcomes and funding sources. As we will see, however, difficulties arise in the operability of this concept for prospective biobanks. (Consent issues can of course also occur in respect of existing collections; especially where the original purpose was not for research. This raises questions regarding waiver of consent and is beyond the scope of this chapter.)

Understanding the Context: Why is Consent a Problem for Biobanking?

In order to understand the problem we need to note a number of factors at play which have contributed to the strain on consent principles in the context of biobanks. From the outset, it should be noted that there are sensitivities surrounding the collection of human genetic samples and information, and this is recognized in international statements such as the UNESCO (2003) *Declaration on Genetic Data*. The most significant aspect, however, is the very nature of biobanks which typically involve large-scale, longitudinal population-based research for which data needs to remain at least potentially identifiable for purposes of data linkage. Added to this is the complexity of future undefined projects. Realistically, at the time of initial collection of samples, specific informed consent can only be secured for any known specific use and for storage of that data. At this early collection stage, only broad consent can be obtained for future use in as yet unspecified research projects. This is because it is impossible to provide specific information at this stage about each project's aims, methods, and arrangements with regard to data access. This has given rise to uncertainty about the legal and ethical validity of broad consent to future unspecified research (Australian Law Reform Commission and the Australian Health Ethics Committee (ALRC/AHEC) 2003). Opponents of broad consent claim that the inability to give participants specific details about the research projects that their data will be used for undermines the respect for autonomy that the consent measures are designed to protect (Kaye 2004). It is also argued to be problematic in practice and unduly simplistic to have one-off consent for studies which may run over several decades: people may change their mind about what they do or do not regard as legitimate use (McHale 2004).

Prevailing National and International Ethical Standards

A review of the prevailing ethical standards for consent in the context of biobanking reveals that some form of broad consent is suggested under most governance structures. Internationally, there are a number of statements which acknowledge the importance of genetic research and the fact that specific consent is not always feasible. Some organizations advocate as a solution the anonymization of samples (e.g. United Nations Educational Scientific and Cultural Organisation (UNESCO 2003) *International Declaration on Human Genetic Data*, World Health Organisation (WHO 2004) *Statement on Genetic Databases: Assessing the Benefits and the Impact on Human and Patient Rights*), but in many cases, this is not practicable because of the need to maintain capacity to link data in order to be able to contact subjects in the event that significantly relevant health information comes to light. Others go further and expressly support the concept of broad consent (e.g. Human Genome Organisation (HUGO 2002) *Ethics Committee Statement on Genetic Databases* and more recently, the OECD (2008) *Draft Guidelines for Human Biobanks and Genetic Research Databases* which acknowledge the ethical permissibility of broad consent). This tension between statements highlights the problem of lack of consensus internationally on the regulation of such collections with respect to the issue of consent.

At the supranational and national level in Europe, the prevailing approach adopted in many European guidelines (e.g. Council of Europe (2006), *Recommendation Rec(2006) 4 of the Committee of Ministers to Member States on Research on Biological Materials of Human Origin*; German National Ethics Council (2004), opinion on *Biobanks for Research*) is that broad consent is acceptable for unspecified future research use of samples, provided that the research is subject to ethics review and that participants have the right to withdraw at any time. In the United Kingdom, the *Human Tissue Act 2004* and the Human Tissue Authority (2006), *Code of Practice on Consent* endorse the principle of broad consent, and even go so far as to say consent is not required if the researcher is not able to identify the person, and the research is ethically approved by a research ethics committee. UK Biobank has also supported broad consent (UK Biobank 2006).

In the United States, there have been shifting boundaries; although the NBAC (National Bioethics Advisory Commission 1999) had recommended that research conducted with coded samples is research on human subjects and regulated by the Common Rule, research with coded samples has more recently been redefined as not involving human subjects, and therefore not subject to consent requirements or ethics review (Office for Human Research Protections 2004 and reaffirmed in the 2008 guidance document). These developments help to illustrate that there is a problem in the application of consent, but seek to solve it by sleight of hand and what effectively involves manipulation of the status of the genetic data. Indeed, one might question the ethical standing of a solution the main goal of which is to escape existing regulation so that most genetic research can take place without further surveillance.

Developments in Australia can be traced back to the ALRC/AHEC Inquiry into the Protection of Human Genetic Information which culminated with the *Essentially Yours* Report in 2003. The Inquiry exposed differing views *about* and interpretations *of* the application of consent where people cannot be informed in any detail about the nature of future research, and also concerns that too strict an interpretation would handicap research. Against this background, the *Essentially Yours* Report called for clarification through revision of the National Health and Medical Research Council (NHMRC) *National Statement on Ethical Conduct in Research Involving Humans* to more clearly deal with the issue of consent to future use so as to provide guidance to researchers.

The revised *National Statement on Ethical Conduct in Human Research* (NHMRC 2007) has now clarified the situation. In the relevant chapter it recognizes that, subject to certain safeguards, participants may give their consent for the use of their data in future unspecified research ('unspecified consent') including permission to enter data or tissue into a databank or tissue bank. To put the matter beyond doubt, the statement provides that the necessarily limited information and understanding about research for which unspecified consent is given can still be sufficient and adequate for purposes of consent. This seems a practical and sensible approach which gives due weight to the *substance* of consent rather than preoccupation with its *form*. By endorsing the ethical permissibility of this stance, it answers calls by the ALRC/AHEC Inquiry for clarification of the matter and provision of guidance to researchers and also addresses concerns voiced by researchers that an overly strict approach to consent could seriously impede genetic research.

Overall, the trend internationally seems to be in favour of more flexible consent requirements, particularly at the supranational and national level in Europe, and as noted, some jurisdictions in fact go further, stipulating that neither consent nor human ethics review is required, depending on how the data is characterized.

Key Principles Underlying Consent

In evaluating the argument for specific consent, and considering whether there are alternative options, we need to identify: what it is that is important to protect? What are the key principles underlying consent? Essentially, we need to be able to answer the question: is the specificity of consent in the context of prospective biological collections really necessary to protect the principles and values that consent stands for?

There are a number of reasons why the consent of research subjects is necessary in order for the research to be considered ethically justifiable. These can be summarized down to the need to ensure autonomy, respect for and protection of subjects. A review of the human research ethics literature indicates a very close nexus between the principle of informed consent and respect of personal autonomy. Autonomy involves respect of the self-determination of individual subjects and their right to make decisions about matters relevant to personhood, including their bodily integrity (Beauchamp and Childress 2004). This, in turn, implicates human rights norms including dignity (Caulfield and Brownsword 2005) and respect for persons to ensure that subjects are valued in themselves and not merely used as a means to an end. However, O'Neill (2003) has cautioned against overemphasis on autonomy as the guiding principle in relation to informed consent arguing that it is open to varying interpretations, and contemporary accounts have lost touch with their Kantian origins, often lacking the necessary link with respect for persons. Instead, she argues that we need to take consent seriously because it provides an assurance that patients and others are neither deceived nor coerced. This focus on ensuring that consent is genuine highlights that an important aspect of consent is to protect subjects from harm. O'Neill (2003: 6) articulates her view of how this can best be achieved in terms of access to 'extendable information', and subjects having given 'rescindable consent'. This requires researchers to adopt an honest and transparent approach with prospective subjects giving them accurate and relevant, but not overwhelming amounts of information, so that they can decide whether to participate in light of their own values, and assessment of risk factors, as well as their trust in those seeking consent (Campbell 2007). Underlying O'Neill's arguments is her belief that complete, wholly specific consent is in any event an illusion, and that *more specific* consent is not invariably *better* consent from a qualitative point of view (O'Neill 2003). Rather than focusing on highly detailed and specific consent, the aim should be on achieving *genuine* consent. O'Neill argues that genuine consent is achieved where subjects can control the amount of information that they receive and what they allow to be done. There are strong arguments to be made that it respects the autonomy of subjects more to leave control of how much information they need and want to them, rather than paternalistically assuming that an information deficit undermines their capacity to give consent (Campbell 2007). Thus, from the foregoing, we can distil autonomy, respect and protection as the primary ethical principles underpinning informed consent.

Possible Approaches to Consent in the Context of Biobanks

Looking at possible approaches to consent in the context of biobanks, there is a spectrum of options ranging from specific re-consents in every case at one extreme through to the waiver of participant consent at the other.

Specific re-consent in every case

Without seeking to diminish the importance of full informed consent where it can be achieved, there are compelling reasons why we should not insist on specific consent with regard to samples and information collected for biobanks. In particular, the impracticability of such a requirement would be unduly burdensome for both researchers and participants and the process itself may in fact operate as a disincentive to research participation and this is recognized in various national and international reports and statements (e.g. Academy of Medical Sciences 2006). Significantly, there are alternative options available which can avoid these impediments but at the same time uphold the core principles underpinning consent. As will be elaborated below, the model that I am arguing for is in fact a *hybrid* model: informed and specific consent for the collection and storage of samples and information, and *broad* consent for future use for presently unspecified research.

Blanket consent

Broad consent is to be differentiated from 'blanket' consent (*unrestricted* right to use the sample/information in *any* research without any other information). This is too wide to be meaningful and, by comparison, helps to show that broad consent does have clear parameters, namely, studying the role of genetics in disease.

Implied or presumed consent

In some circumstances, in the absence of express consent, agreement can be implied or presumed on the grounds that the person would most likely be willing to give their consent. It could be argued that because only a relatively small proportion of individuals would want to be recontacted for reconsent, consent to future use can be presumed. However, this model of presumed consent is problematic because of the degree to which it undermines respect for the individual participant and the right of the individual to make informed decisions about matters involving him or her.

Waiver of consent by research ethics committees

Waiver of consent may take different forms. For example, consent may be waived subject to anonymization of data; however, it may involve use of identifiable material without consent. It is usually contemplated that the process of waiver of consent is performed by a properly constituted human research ethics committee. The rationale of waiver is that the consent of the individual can be dispensed with in circumstances where there is a compelling public interest in the research proceeding and obtaining consent is impracticable. It is important to recognize in these situations there is no actual consent from the individual concerned. If waiver of consent can be ethically permissible in some circumstances, there is an argument to be made on the grounds of ethical consistency: As Hanson et al. (2006) argue, given that research ethics committees might permit biobank or other research without consent, it seems odd that participants themselves should not be able to give broad consent to use of the material in future unspecified research, which would in any event be subject to human ethics committee review.

The Case for Broad Consent

Returning to the subject of broad consent, the essence of my argument is that the principles which underpin consent are *normally* served by obtaining specific consent where that is possible. Where this is not feasible, reconsent is not necessary in every case to promote these principles. I would argue that they can just as well be advanced through seeking broad consent to future use, provided that there is openness in approach and safeguards to ensure ongoing protection of subjects' interests. Indeed, viewed against the background of this range of consent options outlined above, broad consent is most closely aligned to specific consent, and, it must be remembered, arises in the context where specific consent is sought for the initial collection and storage of the samples and information, so we are not talking about broad consent in a vacuum. This hybrid approach can be supported as a matter of principle, not just on grounds of expediency or public interest, although there is considerable weight of opinion that supports the latter argument (e.g. HUGO 2002).

In making the case for broad consent, it must again be stressed that a more specific consent is not necessarily better quality consent (O'Neill 2003). Although broad consent may be perceived as a less strict requirement than specific consent (Elger and Caplan 2006), it is submitted that the core principles in relation to consent (autonomy, respect, protection) need not be compromised and this form of consent can be as effective in achieving *genuine* consent. Commentators such as Caulfield have maintained that the foundational principle is that the interests of participants must remain paramount (Caulfield 2007; and note Article 5 of the *Declaration of Helsinki* [1964]); I endorse this position and would claim that this hybrid model, if properly implemented, represents a key part of actively respecting the autonomy of research subjects and of safeguarding their interests. Indeed, it can be argued that a detailed and complex consent process may be more burdensome than protective (Chadwick and Berg 2001, O'Neill 2003).

In evaluating the case in support of broad consent, it is helpful, I would suggest, if we try to think through what subjects are being asked to consent to when they are invited to participate in a biobank. The very purposive nature of population biobanks – broadly, a research platform to support genetic research – makes it relatively easy to generalize about the aims and objectives of research that is likely to be undertaken using the samples and information in that biobank. In general terms, all genetic research using biobanks will be to advance the understanding of the role of genes in human health and disease, and the interactions between genes and the environment. If we reflect on what subjects would normally be expected to be informed about when their specific consent is sought (aims, methods, risks and harms etc.), much of this can be dealt with in advance in more generic terms. Whilst this would not be with the same detail as if there was a specific project being described, the question must be asked, are details about these matters really relevant to individuals making an autonomous choice or in other respects, safeguarding the interests of participants? On key issues to do with demands, risks, harms, inconvenience and discomfort, the relevant information in respect of these matters would, to a large extent, be covered in connection with the collection and storage of the samples and information for which a *specific* consent could be obtained, and these risks are in any event low. The broader risks such as unauthorized release of data, possible psychosocial and/or economic harms, or discrimination would potentially arise with most such projects involving the use of human genetic samples and information and therefore could be stated in general terms; the risk for participants is low, provided adequate safeguards are in place.

Thus, 'broad consent', can be elucidated and specified to some degree due to the relatively homogenous nature of genetic research – whilst participants may not have the specifics of each research project, they have enough information to understand the *general* nature of what the research is about. This is the view promoted in the Australian *National Statement* 2007: that the necessarily limited information and understanding about research for which unspecified consent is given can still be sufficient and adequate for purposes of consent.

It is also important to note that while research projects remain unspecified at the time consent is obtained for biobank participation, over time, detailed and specific information will become available. Therefore, with open communication channels, subjects can be kept informed if they wish, of prospective research projects, allowing them an opportunity to withdraw their participation if they object to any project. Rather than conceiving of consent as a single event, which is particularly unsuitable for biobanks which envisage multiple uses of samples for different projects over an extended time period, consent can be seen as a *continuing process* which the subject controls. In a sense, this is nothing new: participants have always had the right to withdraw their participation from research which in turn imposes obligations on the researchers to ensure that subjects have the means to communicate with them on an ongoing basis. Accordingly, it is contended that broad consent can be effective and meaningful. When combined with a specific consent for collection and storage of samples, it fulfils the key protective role of guarding against harm. Further, it promotes the idea of consent as a continuing *process*, which the subject controls by means of information accessible to them. This is highlighted by the approach of UK Biobank which invites participants to 'consent to participation in a biobank' (UK Biobank 2006). It also respects the autonomy of subjects; indeed, it can be argued that to deny people the opportunity to give broad consent on the grounds that it cannot be informed, is a paternalistic approach which defeats the very autonomy informed consent is meant to protect (Campbell 2007). However, in order to be effective, it is essential, that this approach operates within a well regulated and protective environment augmented by a strong governance regime: indeed, it would be a mistake to rely on consent as the sole basis of protecting research participants. Even at its best, specific consent cannot do all the work of protecting individuals (O'Neill 2003), especially in this sensitive context involving genetic data which is seen by many as having a special status (as, for example, recognized in the UNESCO [2003] *Declaration on Genetic Data*).

Embedding Safeguards for Broad Consent

Turning to consider the safeguards required for effective broad consent: Firstly, a non-negotiable core requirement is ethical review and oversight – researchers need to present to the ethics committee details of the research project as well as a record of the broad consent given by participants, including any restrictions so that the ethics committee can check for itself the issue of scope of the consent in relation to the proposed project. It is conceivable that in some circumstances, even if broad consent is initially obtained, that consent may need to be reaffirmed if there was a major modification to the research direction of the biobank so that it is of a fundamentally different nature to that initially proposed, for example, with a new biobanking partner or significantly different aims. In these circumstances, an ethics committee might justifiably require that subjects be recontacted for specific consent before research can proceed, or alternatively, consider waiver of consent.

Secondly, there should, so far as possible, be full disclosure to participants, including areas of likely future research. This would facilitate subjects making informed decisions about their ongoing research participation, empowering them to identify changed circumstances that may lead them to exercise their right to withdraw.

Further, there are a number of strategies that can be taken to help promote ongoing active consent: after the initial collection of samples and information, it is important that open lines of communication are maintained so that subjects are kept informed of research directions and possible planned uses of samples and information in the future. This could, for example, be achieved through project websites, periodically updated as well as regular email updates or newsletters, spelling out research plans and planned uses of data. In this way, one can give control to subjects about how much information they want, and those that want to follow up and obtain details about specific projects can do so. It is, therefore, important that plans for ongoing provision of communication with subjects be carefully articulated at the time broad consent to future unspecified use in research is sought.

Linked with this is the participants' right to withdraw consent and have their samples and information removed from the study (in so far as is feasible). This creates a corresponding obligation on researchers to ensure that participants have the means to communicate with them to notify their wish to withdraw. It is, therefore, incumbent on biobanks to be proactive and maintain visibility within the community.

Expectations and Interests of Research Participants: The Role of Public Opinion Data

Also of relevance in this context are the expectations and needs of research subjects. There appears to be a tendency to attribute to participants the desire to have full control, and it is assumed that this means that they are not comfortable with the notion of consent to unspecified research and would prefer to be contacted for specific consent for each project. However, this assumption is not necessarily well-founded; rather, it seems, that what subjects want is transparency and honesty in conduct, and that the preservation of trust is critical; this would entail full disclosure of present intentions (including commercialization potential) (Caulfield 2007, Caulfield et al. 2007) and assurances about appropriate safeguards to protect subjects' interests, the confidentiality of their data, and that it will not be misused (Hoeyer et al. 2004).

Whilst the limitations of reliance on public opinion data need to be acknowledged, it is submitted that it is justifiable to have regard to such data in formulating law and policy, as legal principles do not operate in a vacuum, and must be responsive to the public interest. There is a range of data, some of it conflicting, probably reflecting variation in how questions are framed (Caulfield 2007). Certainly, the public is very accepting of and supportive towards genetic research in general (Wendler 2006), including biobanks, and many indicate their willingness to participate (Kettis-Linblad et al. 2006, McQuillan et al. 2003, Secko et al. 2009, Nicol and Critchley this volume). Further, there is some empirical data which suggests that research participants support this broader approach to consent, provided that appropriate safeguards are in place (Wendler 2006). For example, some participants eligible for UK Biobank have endorsed broad consent because they do not want to be recontacted repeatedly; indeed repeated consent may be seen not only as impracticable but also as intrusive from the research participant's perspective and may in fact operate as a deterrent (Opinion Leader Research 2003). From other research, there is some evidence of consent 'fatigue' and of people not wanting extensive information. For example, Swedish research has found that consent is not the issue perceived as most important by respondents; rather, they were concerned about the confidentiality of medical records (Hoeyer et al. 2004). These findings are consistent with the altruistic purpose behind donations of human genetic samples and information. Inevitably, however, there is some diversity of views in relation to consent: Whilst some people may be 'health-information altruists' (Kohane and Altman 2005), there are others who would undoubtedly prefer to be recontacted for specific consent in connection with a new project rather than give consent to future unspecified use (Caulfield 2007). Although this may be their clear preference if given the option of specific consent, arguably, these same people may be comfortable to trust researchers if it is explained to them that reconsent is not feasible for large-scale biobanks, and that the only way such research can proceed is with broad consent to future as yet unspecified research. We must remember that ultimately, research participation is voluntary, and if prospective participants are not comfortable with the terms for participation, they are free to decline involvement.

Thus, an approach is needed which maximizes the opportunity for research subjects to participate and engage in biobanking and for genuine partnerships between biobanks and their participants (Campbell 2007). In order to promote trust and facilitate research participation, it is important at this nascent stage of biobank development that these resource infrastructures are appropriately set up with a rigorous ethical framework and pursuant to an evidence-based approach which takes account of the expectations and interests of research subjects. While existing data generally supports the trend in favour of broad consent, further research is required to more clearly define the parameters of what subjects or potential subjects regard as acceptable, and also what safeguards they attach significance to in ensuring that their interests are protected.

Conclusion

If one accepts, at a practical level, that specific consent for each research project is not possible in respect of biobanks and that there are important public interests at stake, the conclusion is inescapable that some ground needs to be given on the issue of consent. Attention must therefore turn to the manner in which consent can be successfully operationalized in this context without compromising the core principles it seeks to protect. However, any modification of requirements for consent cannot simply be justified on considerations of expediency but must be defensible on the basis of a principled approach which preserves the central principles and values that consent represents. I have argued that consent needs to be reconceptualized in the context of biobanks, but that the key principles that consent is intended to serve, namely autonomy, respect and protection must be upheld. I have asserted that this can be achieved through a hybrid model involving broad consent to future use given that specific consent is obtained at the time of collection and storage of samples, and that the integrity of these principles is protected by safeguards.

In sum, I have sought to argue that broad consent should not be viewed as an inherently inferior option: With appropriate safeguards, it arguably better respects the autonomy of willing participants and avoids paternalistic assumptions regarding consent. Importantly, I am not proposing that the public interest in biobanking should simply trump individual interests; it is too simplistic to see this solely as a conflict between public and private interests. As I suggested at the outset, there is an important public interest in ensuring appropriate protection and respect for the rights of individual research subjects because preserving community trust is fundamental to the facilitation of such research. This chapter has tried to demonstrate that the hybrid consent model proposed promotes individual interests and gives them the paramountcy that they deserve.

Bibliography

Academy of Medical Sciences. 2006. *Personal Data for Public Good: Using Health Information in Medical Research.*

Australian Law Reform Commission and the Australian Health Ethics Committee (ALRC/AHEC). 2003. *Essentially Yours*. Canberra.

Beauchamp, T. and Childress J. 2004. *Principles of Biomedical Ethics*. 5th Edition. Oxford: Oxford University Press.

Campbell, A. 2007. The Ethical Challenges of Genetic Databases: Safeguarding Altruism and Trust. *Kings Law Journal*, 18, 227-245.

Cambon-Thomsen, A. 2004. The Social and Ethical Issues of Post-Genomic Human Biobanks. *Nature Reviews: Genetics*, 5, 866-873.

Cambon-Thomsen, A., Rial-Sebbag, E. and Knoppers, B. 2007. Trends in Ethical and Legal Frameworks for the Use of Human Biobanks. *European Respiratory Journal*, 30, 373-382.

Caulfield, T. 2007. Biobanks and Blanket Consent: The Proper Place of the Public Good and Public Perception Rationales. *Kings Law Journal*, 18, 209-226.

Caulfield, T., Brown R. and Meslin, E. 2007. Challenging a Well Established Consent Norm? One Time Consent for Biobank Research. *Journal of International Biotechnology Law*, 4, 69-74.

Caulfield, T. and Brownsword, R. 2005. Human Dignity: What does it Mean in the Biotechnology Era? *Nature Review Genetics*, 7, 72-76.

Chadwick, R., and Berg, K. 2001. Solidarity and Equity: New Ethical Framework for Genetic Databases. *Nature Review Genetics*, 2, 318-321.

Council of Europe. 2006. *Recommendation Rec(2006) 4 of the Committee of Ministers to Member States on Research on Biological Materials of Human Origin.*

Elger, B., and Caplan, A. 2006. Consent and Anonymization in Research Involving Biobanks: Differing Terms and Norms Present Serious Barriers to an International Framework. *EMBO Reports*, 7, 661-666.

Hanson, M., Dillner, J., Bartram, C., Carlson, J. and Helgesson, G. 2006. Should Donors be Allowed to Give Broad Consent to Future Biobanks Research? *The Lancet Oncology*, 7, 266-269.

Hoeyer, K., Olofsson, B., Mjörndal, T. and Lynöe, N. 2004. Informed Consent and Biobanks: A Population-based Study of Attitudes Towards Tissue Donation for Genetic Research. *Scandinavian Journal of Public Health*, 32, 224-229.

Human Genome Organisation (HUGO). 2002. *Statement on Human Genetic Databases.*

Human Tissue Authority. 2006. *Code of Practice on Consent.*

Kaye, J. 2004. Abandoning Informed Consent: The Case of Genetic Research in Population Collections, in *Genetic Databases: Socio-Ethical Issues in the Collection and Use of DNA*, edited by R. Tutton and O. Corrigan. London/New York: Routledge, 131.

Kegley, J. 2004. Challenges to Informed Consent: New Developments in Biomedical Research and Healthcare May Mark the End of the Traditional Concept of Informed Consent. *EMBO Reports*, 5, 832-836.

Kettis-Linblad, A., Ring L., Viberth, E. and Hansson, M. 2006. Genetic Research and Donation of Tissue Samples to Biobanks: What do Potential Samples Donors in the Swedish General Public Think? *European Journal of Public Health*, 16, 433-440.

Kohane, I., and Altman, R. 2005. Health-Information Altruists – A Potentially Critical Resource. *New England Journal of Medicine*, 353, 2074-2077.

McHale, J. 2004. Regulating Genetic Databases: Some Legal and Ethical Issues. *Medical Law Review*, 12, 70-96.

McQuillan, G., Porter, K., Agelli, M. and Kington. R. 2003. Consent for Genetic Research in a General Population: The NHANES Experience. *Genetics in Medicine*, 5, 35-42.

National Bioethics Advisory Commission. 1999. *Research Involving Human Biological Materials: Ethical Issues and Policy Guidance*.

National Ethics Council. 2004. *Biobanks for Research (Opinion)*. [Online] Available at: http://www.ethikrat.org/_english/publications/Opinion_Biobanks-for-research.pdf.

National Health and Medical Research Council (NHMRC). 2007. *National Statement on Ethical Conduct in Human Research*.

Nuremberg Code (1947).

OECD. 2008. *Draft Guidelines for Human Biobanks and Genetic Research Databases*.

Office for Human Research Protections. 2004. *Guidance on Research Involving Coded Private Information or Biological Specimens*.

Office for Human Research Protections. 2008. *Guidance on Research Involving Coded Private Information or Biological Specimens*.

O'Neill, O. 2003. Some Limits of Informed Consent. *Journal of Medical Ethics*, 29, 4-7.

Opinion Leader Research. 2003. *Summary of the UK Biobank Consultation on the Ethics and Governance Framework*.

Otlowski, M. 2008. Creating a Facilitative Environment for Genetic Research: The Central Role of Public Trust, in *The Brave New World of Health*, edited by B. Bennett, T. Carney and I. Karpin. Sydney: Federation Press, 128-143.

Secko, D., Pretor, N., Niemeyer, S., and Burgess, M. 2009. Informed Consent in Biobanks Research: A Deliberative Approach to the Debate. *Social Sciences and Medicine*.

UNESCO Declaration on Genetic Data (2003).

UK Biobank. 2006. Ethics and Governance Framework, Version 2.0.

Wendler, D. 2006. One-Time General Consent for Research on Biological Samples. *British Medical Journal*, 332, 544-547.

WHO. 2004. *Statement on Genetic Databases: Assessing the Benefits and the Impact on Human and Patient Rights*.

World Medical Association. 2000. *The Revised Declaration of Helsinki: Interpreting and Implement Ethical Principles in Biomedical Research*.

Chapter 6
Consent by Research Ethics Committees: The New Law on Biomedical Research in Spain

Antonio Casado da Rocha and Ismael Etxeberria Agiriano

Enacted in July 2007, the Law on Biomedical Research (*Ley de Investigación Biomédica*, Ley 14/2007) provides a specific and novel regime for biobanks in Spain. This is not a specific law for biobanks, but a more general one attempting to strike the difficult balance between scientific advancement and traditional ethical values in the regulation of human biological samples, their collection and use in diagnosis and biomedical research. It also seeks to promote the interests of research institutions and companies. In doing so, the new legal regime for biobanks in Spain relies heavily on the decision-making by research ethics committees. How this will work is currently unclear, as key concepts in the legislation are not yet defined. Such ambiguity makes room for many possible outcomes, and the new law might eventually set in place something very similar to an 'open consent' or 'consent waiver' model, far away from the traditional demand of informed consent.[1] To succeed, this movement from the traditional requirements of informed consent towards a 'governance-by-committee' model will need, in the short term, recommendations and new research ethics committees to take on the role of making decisions on behalf of participants. In this chapter, we discuss the Spanish context, the way the new legislation deals with informed consent, and our concerns about this new development.

The Growth in Biomedical Research in Spain

Biomedical research in Spain has undergone significant growth in recent years. According to the bibliometric map for biomedicine and health sciences in Spain (Méndez-Vásquez et al. 2008: 248) in the period 1996-2004 research activity in this field grew by 8.9 per cent in the number of documents and by 22.5 per cent in the number of citations, compared to the period 1994-2002. The field of

1 The 'consent waiver' and the 'open consent' models are not to be confused; they represent two possible but different outcomes of the trend we identify and briefly describe here (see also Nys and Fobelets 2008: 177).

biomedicine occupies the second place in number of publications (40.3 per cent of total publications in Spain), but is first in terms of the number of citations (51.3 per cent of total citations). While Spanish biotechnology is being concentrated in the human health sector, there is only one associate biobank member of P^3G (*Genoma España*) from Spain, and there has been less public consultation about these issues than in other countries such as the UK. The critical mass of researchers in Spain is relatively low, constituting only small and fragmented groups, even though socioeconomic and health status indicators in life expectancy, vaccine coverage, or maternal and infant mortality show significant growth over previous years (Durán et al. 2009: 82). The new Law on Biomedical Research (or Law 14/2007) therefore is a response to the growth in this new field but also suggests that it is anticipated that this new field will continue to expand. The law provides a legal framework for this emerging area and is one of the few examples in the world where a specific legislation has been developed for biobanks.

Informed Consent

Informed consent is probably the most studied and controversial issue in the literature on the ethical, legal and social aspects of biobanks, and the debate is far from being resolved. Some researchers feel that the vast array of guidelines set up to control the way samples are collected and used in biobanks could create confusion and thus hamper science. Indeed, this so-called 'biobank confusion' (MIP 2007) might have resulted in unnecessary hurdles for genome-based research, particularly when samples are shared between different institutions in several countries (Zika et al. 2008). Arguments in favour of harmonizing the terminology about anonymity and allowing general consent have been presented (Elger and Caplan 2006: 665) and the new Spanish legislation follows closely such attempts. In addition, some biobanks (specifically, tumour banks) have tried to harmonize their technical procedures and ethical requirements, aiming towards a cooperative and coordinated network of networks in Spain (Morente et al. 2007).

As such, biobanks are not a novel thing in Spain or elsewhere, but recent avenues opening in genomic research have renewed interest in establishing new biobanks, as well as increasing access to existing collections. As is well known, the ethical issues surrounding biobanks attracted world attention in the late 1990s, when the debate was focused on the Icelandic case of deCODE Genetics. This case has become the exemplar of how tangled the relationship can be between conflicting interests in this field (Árnason 2004). Since then, much of the 'biobank confusion' comes from the attempt to apply traditional regulatory frameworks to new technologies. After the Declaration of Helsinki, informed consent requires that individuals are given information about the research, but 'this is often impossible to do at the time data and samples are collected, since biobanks are designed to be used by many researchers and for many projects well into the future' (Kaye 2004: 103).

The result is that informed consent remains in most declarations at the global level the standard ethical requirement for medical research, but at the same time the new circumstances of biobanks are thought by many to be sufficiently different to require a different ethical and legal standard. In general, it seems that the views of both experts and potential donors are becoming more favourable to 'open consent' or 'one-time general consent' models (Wendler 2006), in which informed consent must be required only for the initial collection of samples and data:

> at this occasion, donors should have the choice to limit their participation to already known research projects about which they receive detailed information, or to provide, in addition, a general consent to future research projects the precise scope of which is still unknown. (Andorno 2007: 37)

However, many difficult questions remain, particularly around international collaborative projects. The pooling of biosamples in international studies increases their potential scientific value, but the multiplicity of guidelines in local legislation creates uncertainty and raises new questions. In particular, what is to be done with previously collected samples? What are the appropriate information and consent procedures in this context of international cooperation? The new Spanish legislation goes some way to answering these questions.

The Spanish Legal Framework

Biomedical data in Spain are not subjected to a unique, specific piece of legislation. The collection and use of personal data and healthcare information (including genetic data), as well as informed consent procedures, are regulated by a cluster of legal instruments – such as the Organic Law 15/1999 on the Protection of Personal Data; the Law 41/2002 on Patients' Freedom, Rights and Duties on Information and Clinical Documentation; and the European Convention on Human Rights and Biomedicine (CoE 1997), which provides a more general framework.

The Law on Biomedical Research is the latest piece in this jigsaw puzzle and has significantly changed the legal landscape. Its aim is to encourage biomedical research while ensuring the highest possible level of ethical and legal guarantees to protect the general population. As a matter of fact, it was the result of a comprehensive legislative effort, encompassing the promotion, planning, evaluation, coordination and mobility mechanisms involved in all biomedical research, both in the private and public sectors, and its consequences affect a population of more than 45 million people. This was an ambitious project and as such it was bound to stir some controversy.

Controversial aspects of the law on biomedical research

The Act included a number of new initiatives, in addition to the new biobank legislation, within the area of medical research that have created controversy, such as the creation of a Spanish Bioethics Committee in 2007. This Committee did not become fully operational until 2009 because of the creation of a new Ministry for Science and Innovation which followed the elections in 2008, as this Ministry assumed the task of promoting and coordinating most of biomedical research in Spain. Initially, the Law on Biomedical Research attracted public attention because it was thought to regulate the genetic selection of embryos for therapeutic purposes. As a matter of fact, there is no provision in the law regulating that aspect, but it covers many related, important issues such as: the regulation of clinical trials and other health research in humans with invasive procedures; the donation and use of embryos and foetuses – including their cells, tissues and organs – for research and clinical applications; genetic tests and the use of personal data in the health care sector.

The drafting process of the Law on Biomedical Research took into account and sought accordance with international norms and standards which Spain must follow. These were the above mentioned European Convention on Human Rights and Biomedicine (also called the Oviedo Convention, CoE 1997); the Directive 2004/23/EC on setting standards for the donation and use of human tissues and cells (EP 2004); as well as the Additional Protocol to the Convention on Human Rights and Biomedicine concerning Biomedical Research (CoE 2005), which at the time of the passing of the Law remained unsigned by Spain. The new act also supplements the Law on Assisted Reproduction Techniques (Ley 14/2006), which prohibits 'reproductive cloning', as Spain signed the Additional Protocol to the Oviedo Convention on the Prohibition of Cloning Human Beings. The drafting process also took into account the Council of Europe's Recommendation Rec(2006)4 of the Committee of Ministers to member states on research on biological materials of human origin, adopted on 15 March 2006.

The structure of the act

Even if it did not attract much public discussion at the time, one of the most novel aspects of this law concerns the creation, in a specific chapter, of a regime for the regulation of biobanks within the wider framework of biomedical research. Structured in eight main sections or 'Titles', the law devotes Title V to the regulation of human biological samples, their storage and use in medical diagnosis and research, and Chapter IV within this Title V contains specific rules regarding biobanks. The Act defines a biobank in article 3 as:

> a non-profit institution, public or private, storing a collection of biological samples for purposes of biomedical diagnosis or research, conceived of as a technical unit with common criteria of quality, organization and purpose.

Articles 58-62 regulate how samples are to be obtained, preserved, used and transferred, and the issue of the regulation of genuine informed consent appears again. In the Preamble, that gives guidance on the spirit behind the legislation, it says that the intention of the law is to be focused on the donor's consent and the information needed in order to secure it. The law is described as an effort towards a 'flexible, middle way' in between open (or generic) and specific (or traditional) types of consent. Thus, the initial act of consent might include consenting to further, related but unspecified uses of the samples. The law allows for specific acts of consent to include consenting to the use of data and samples in other research projects, 'related to the one initially proposed', by the same team or another one (Art. 60.2). The 'degree of relationship' between the two projects remains unspecified and open to interpretation by the relevant body, usually the Research Ethics Committee supervising the biobank.

In addition, some of the 'Dispositions' (that is, practical enforceable provisions placed outside specific articles at the end of the legal text) are of particular relevance to the governance of biobanks. For instance, the Second Transitory Disposition concerns what is to be done with samples stored before the passing of this law, and the Third Final Disposition empowers the Government to regulate the transfer of samples abroad.

Project relatedness

The notion of 'project relatedness' is undefined in the legislation, and to date there has been no case law on this point. There seems to be at least two ways of understanding 'project relatedness', which is to take either a wide or a strict interpretation. In a strict sense, individual research projects are related when they represent 'lines of enquiry' pursued by the same team, with the core theory and the body of previous research changing only slightly from study to study, and much of the prior work being reused in each new line of research. In a wide interpretation, and taking into account the speed of scientific development in the area of genetics, and the vast spectrum of potential research hypotheses that may arise and be addressed by biobanks (Shickle 2006), there will be a range of possible uses and no easy way to predict what the range will be.

Research ethics committees will have to decide the issue of 'project relatedness' and will have to produce appropriate guidelines and criteria. However, no two ethics reviews are the same, and a large variation in results is to be expected. If the requirement of project relatedness is generally understood in the wide, more inclusive sense, what the Spanish law puts in place would virtually be a single-act, open-consent, non-commercial model, with opting-out and other procedural safeguards. However, much still needs to be decided, and we will have to 'wait and see' the decisions of the research ethics committees and the courts as the law evolves and is put into practice. As of 2008, there is not a national body that will produce guidelines for all the other research ethics committees. Still, the draft of a forthcoming Law on Science, which is expected to be passed in 2009, includes an article creating a Spanish Research Ethics Committee whose mission includes the production of such recommendations and guidelines.

Research on existing samples

The new Spanish model also uses the 'one-time general consent' approach to the regulation of the samples already collected and stored in existing biobanks. According to the Law on Biomedical Research, consent can be obtained upon the collection of the sample, or at a subsequent time (Art. 60.1). In section IV of the Preamble it is explained that the law had to make provisions concerning 'biological samples obtained for any purpose before the passing of this law, so as to make possible their use for research, while at the same time protecting the donors' interests'.

The Second Transitional Provision of the law lays out the situations when existing samples may be used for biomedical research. These are when:

- the donor has consented to this; or
- they have been anonymized; or
- the process of obtaining consent entails an 'unreasonable effort' (this key concept is defined in Art. 3.i as 'a disproportionate amount of time, work or other expenses'); or
- the donor is dead or cannot be located.

The use of the samples for all of these situations only requires the approval of a Research Ethics Committee, which will examine whether the following conditions are met: (*a*) the research is of general interest; (*b*) lack of data would make research impossible or less effective; (*c*) there is not an explicit objection to it; and (*d*) the confidentiality of personal data is guaranteed.

These exemptions from the principle of informed consent are far wider for samples collected and stored after the passing of this law (3 July 2007). The general rule is that informed consent must be obtained before a sample is collected; however the law provides a number of exceptions. If obtaining consent for the new use was not feasible or entailed an 'unreasonable effort', the requirement of consent may be waived. It does not matter what purpose the sample was originally collected for, and it needs not be anonymized (that is, the donor's identity can eventually be traced). Exceptionally, says the law, use of the samples requires only the approval by a Research Ethics Committee, which will take into account whether conditions *a-d* above are met, and whether (*e*) research will be carried out by the same institution that initially obtained the sample (Art. 58.2). In addition, once the sample is stored, the law allows its use in other biomedical research projects with no restriction other than those established by the scientific and ethics committees of the biobank, which will grant permission provided that the projects are of a scientific nature, and the request of samples includes information about the ends of the research, and declare that they are not to be used for any other purpose (Art. 69). This gives quite a broad margin to the sharing of samples between biobanks.

Sharing of samples

The import and export of human biological material from, and to, other countries is an issue that is not covered directly by the Law on Biomedical Research. There is a Third Final Disposition in which it grants the Government 'power to dictate as many dispositions as necessary to develop and execute' this act, in order to 'establish the internal, inter- and extra-community regulations on exchange and circulation of biological material of human origin for research'. At the time of writing this chapter (October 2008), no law or decree has been enacted to that purpose. One of the leading research institutions in Spain, the Health Institute Carlos III (Madrid), prepared a draft for disposition in late 2007, but this draft is still being rewritten after receiving feedback from several stakeholders and Ministries. The creation of the Ministry for Science and Innovation which followed the elections in 2008 added to this delay.

In conclusion, approval by a Research Ethics Committee is generally compulsory in order to obtain, transfer and use biobank samples, especially when samples from deceased persons are involved, or when samples are to be used in research unrelated to that for which the sample was obtained (Art. 62). In this sense, despite what the lawmaker says about informed consent being the general rule, the law makes it possible to talk about a 'consent waiver' model, in which most decisions concerning permission to use already collected samples are to be taken by an overseeing third party.

Putting much of the decision making in the hands of the committees associated with each biobank, suggests that new research ethics committees will be needed or existing ones will need their capabilities greatly expanded. Specific guidelines will also have to be developed so that they can properly perform all the functions assigned to them by the Law on Biomedical Research. For instance, the Ethics Committee of the Research Institute for Rare Diseases (*Instituto de Investigación de Enfermedades Raras*) – a centre belonging to the Health Institute Carlos III – has issued a set of recommendations taking into account the requirements of the new Spanish law (IIER 2007). Although this is not the first time that proposals concerning the collection and use of samples have been issued in the Spanish literature (see Romeo-Casabona 2002: 170-3), such explanatory documents need to be distributed more widely in order to reach all stakeholders and thus be more effective.[2]

2 This is not an exceptional situation, as in many countries it is up to Research ethics committees to deal with the tough decisions. For instance, Auray-Blais and Patenaude (2006) propose a model for the management of biobanks in biomedical research in Canada where a medical archivist plays the pivotal role as a data-protection officer. This model had

Conclusion

In this chapter we have identified and briefly discussed a case in which there has been a significant move towards more flexible approaches to the traditional requirements of informed consent, which has made its way into legislation. In a review of the existing literature (Casado and Seoane 2008), most authors reject a 'blanket' consent, in which individuals would authorize, once and for all, *any* type of research with their samples. With the general aim of eliminating unnecessary hurdles to research, an array of alternatives ('open', 'single', 'one-time', 'future', 'general', 'broad') is proposed as a middle-ground between specific and blanket consent, while the right to withdraw from research is considered of the utmost importance by nearly all authors.[3] The success of this trend depends upon how adequately and comprehensively it is reflected by governance mechanisms at the national level. As we have shown in the Spanish case, ethical review boards will play a key role in this new scenario, as they have the final word concerning whether new acts of consent are sought in order to use samples already stored in a biobank. Other countries might follow this trend.[4]

It is clear that, after the enactment of the Law on Biomedical Research, interpretation of the legal situation in Spain depends upon the meaning given to key concepts such as 'project relatedness'. Although according to the law, this concept is not completely open, its ambiguity makes room for many possible outcomes; thus the new Spanish law might eventually set in place something very similar to an 'open consent' or 'consent waiver' model, far away from the traditional demand of informed consent. Of course, we believe that such traditional requirements should be evolved and adapted to new circumstances. There are good reasons to do so: for the sake of public health, scientific progress and efficiency in the distribution of research resources. Our concern is that they should be articulated as exceptions to the principle of informed consent. This new law subscribes to the traditional theory of informed consent as a general rule for all biomedical research, but then allows for so many exceptions that such consent might well disappear as an actual practice. The Act also says that social or scientific interests can never override the welfare of individual participants, but if we do not want the partnership between researchers and participants to be asymmetrical and potentially abusive, accountability, transparency and proper monitoring remain crucial conditions for governance in Spain as elsewhere. Time and further work will be necessary in

---

to reduce the burden placed on the committee responsible for the evaluation of genetics projects and, at the same time, maximize the protection of research participants.

3 Besides the references in Casado and Seoane (2008) and Zika et al. (2008), see Ursin (2008), Brand and Probst-Hensch (2007).

4 In Italy, the National Committee for Biosecurity and Biotechnology has issued a set of guidelines for the creation and validation of biobanks. They define a biobank in terms similar to those of Spanish Law, and follow the aforementioned trend towards allowing the reutilization of samples (GI 2006).

order to assess the promises of the Spanish Law on Biomedical Research, both in scientific and normative terms.

Acknowledgments

This chapter is partially based on previously published work (Casado and Seoane 2008) and has benefited from discussions with the participants at the international conference 'Governing Biobanks – What are the challenges?' (Oxford, 24-26 June 2008). Funding was provided by a Spanish Government research project 'Autonomy in Bioethics and Biomedical Research' (*El concepto de autonomía en bioética e investigación biomédica*, FFI2008-06348-C02-02/FISO).

Bibliography

Andorno, R. 2007. Population Genetic Databases: A New Challenge to Human Rights, in *Ethics and Law of Intellectual Property. Current Problems in Politics, Science and Technology*, edited by C. Lenk, N. Hoppe and R. Andorno. Aldershot: Ashgate, 27-45.

Árnason, V. 2004. Coding and Consent: Moral Challenges of the Database Project in Iceland. *Bioethics*, 18(1), 27-49.

Auray-Blais, C. and Patenaude, J. 2006. A Biobank Management Model Applicable to Biomedical Research. *BMC Medical Ethics*, 7(4).

Brand, A. and Probst-Hensch, N. 2007. Biobanking for Epidemiological Research and Public Health. *Pathobiology*, 74(4), 227-238.

Casado, A. and Seoane, J.A. 2008. Alternative Consent Models for Biobanks: The New Spanish Law on Biomedical Research. *Bioethics*, 22, 440-447.

CoE Council of Europe. 1997. *Convention for the Protection of Human Rights and Dignity of the Human Being with Regard to the Application of Biology and Medicine: Convention on Human Rights and Biomedicine.* [Online] Available at: http://conventions.coe.int/Treaty/en/Treaties/Html/164.htm [accessed 19 March 2009].

CoE Council of Europe. 2005. *Additional Protocol to the Convention on Human Rights and Biomedicine, Concerning Biomedical Research.* [Online] Available at: http://conventions.coe.int/Treaty/en/Treaties/Html/195.htm [accessed 3 April 2009].

Durán, A., Lara, J.L. and Waveren, M. 2009. Spain: Health System Review. *Health Systems in Transition*, 8(4). [Online] Available at: http://www.euro.who.int/Document/E89491.pdf [accessed 17 March 2009].

Elger, B.S. and Caplan, A.L. 2006. Consent and Anonymization in Research Involving Biobanks. *EMBO Reports*, 7(7), 661-666.

EP European Parliament. 2004. *Directive 2004/23/EC of the European Parliament and of the Council of 31 March 2004 on Setting Standards of Quality and Safety for the Donation, Procurement, Testing, Processing, Preservation, Storage and Distribution of Human Tissues and Cells*. [Online] Available at: http://eur-lex.europa.eu/LexUriServ/LexUriServ.do?uri=CELEX:32004L0023:EN:HTML [accessed 19 March 2009].

GI Governo Italiano, Comitato Nazionale per la Biosicurezza e le Biotecnologie. 2006. *Linee guida per l'istituzione e l'accreditamento delle biobanche* [*Guidelines for the creation and validation of biobanks*]. [Online] Available at: http://www.governo.it/biotecnologie/documenti/7.biobanche.pdf [accessed 24 March 2009].

IIER Comité de Ética del Instituto de Investigación de Enfermedades Raras. 2007. Recomendaciones sobre los aspectos éticos de las colecciones de muestras y bancos de materiales humanos con fines de investigación biomédica [Recommendations on the Ethical Aspects of Sample Collections and Human Biobanks for Biomedical Research Purposes]. *Revista Española de Salud Pública*, 81(2), 95-111.

Kaye, J. 2004. Broad Consent – the Only Option for Population Genetic Databases? in *Blood and Data – Ethical, Legal and Social Aspects of Human Genetic Databases*, edited by G. Árnason, S. Nordal and V. Árnason. Reykjavik: University of Iceland Press and Centre for Ethics, 103-109.

Ley 14/2006, de 26 de mayo, sobre técnicas de reproducción humana asistida. Madrid: BOE. [Online] Available at: http://www.boe.es/boe/dias/2006/05/27/pdfs/A19947-19956.pdf [accessed 19 March 2009].

Ley 14/2007, de 3 de julio, de Investigación Biomédica [*Law on Biomedical Research 14/2007*]. Madrid: BOE. [Online] Available at: http://www.boe.es/boe/dias/2007/07/04/pdfs/A28826-28848.pdf [accessed 17 March 2009]. An unofficial English translation is available at: http://www.catedraderechoygenomahumano.es [accessed 17 March 2009].

Méndez-Vásquez, R.I. et al. 2008. Mapa Bibliométrico de España 1996-2004: Biomedicina y Ciencias de la Salud. *Medicina Clínica*, 130(7), 246-253.

MIP Metropolis International Group. 2007. Biobank Confusion Cleared up by New Guidelines. *Laboratory News Online*. [Online] Available at: http://www.labnews.co.uk/laboratory_article.php/2401/2/biobank-confusion-cleared-up-by-new-guidelines [accessed 17 March 2009].

Morente, M.M., Alava, E. and Fernandez, P.L. 2007. Tumour Banking: The Spanish Design. *Pathobiology*, 74(4), 245-50.

Nys, H. and Fobelets G. 2008. The Regulation of Biobanks in Spain. *Law and the Human Genome Review*, 29, 169-188.

Romeo-Casabona, C.M. 2002. *Los genes y sus leyes* [*Genes and their Laws*]. Bilbao/Granada: Cátedra Interuniversitaria de Derecho y Genoma Humano/Comares.

Shickle, D. 2006. The Consent Problem within DNA Biobanks. *Studies in the History and Philosophy of Biology and Biomedical Sciences*, 37(3), 503-519.

Ursin, L. 2008. Biobank Research and the Right to Privacy. *Theoretical Medicine and Bioethics*, 29(4), 267-285.

Wendler, D. 2006. One-time General Consent for Research on Biological Samples. *British Medical Journal*, 332(7540), 544-547.

Zika, E. et al. 2008. Sample, Data Use and Protection in Biobanking in Europe: Legal Issues. *Pharmacogenomics*, 9, 773-781.

Chapter 7
Addressing the Ethical Objections to Pediatric Biobanks

Kristien Hens and Kris Dierickx

Children are no longer regarded by the law as second-rate citizens, replaceable and not worthy of consideration. Since the United Nations *Convention on the Rights of the Child* (1998), children are now seen as persons deserving of special protection, with the right to participate in the decisions that affect them. This change in attitude has also been seen within medical research where children are no longer regarded as mere objects of experimentation. However, even at the beginning of the twenty first century, there is still a reluctance to involve children in research and it is only relatively recently that specific biobanks have been established that focus solely on children's development and diseases.

This chapter will discuss this change in relation to biobanks by discussing the major ethical questions that arise when using pediatric tissue samples stored in biobanks. We shall use the term 'child' in its meaning of legally incompetent minor under eighteen years, as it is not possible within the scope of this chapter to describe all the nuances and the subsequent ethical issues that may arise for children at different ages. The themes we shall focus on are consent, minimal harm and benefit, as these are most extensively discussed in the literature on pediatric biobanks. We will argue that these key ethical objections to research on children and the establishment of pediatric biobanks are ill-founded. By establishing appropriate procedures that are sensitive to children, these objections can be addressed and allow valuable research to proceed.

Children and Medical Research

The history of children as medical research subjects has been characterized by abuse, with little respect for understanding children's rights and concerns. Lederer and Grodin describe the use of children as research subjects until the mid twentieth century as largely one of child abuse: researchers would experiment with new vaccines on their own children, and institutionalized children in particular were enrolled in sometimes questionable research projects (Lederer and Grodin 1994). The atrocities of Nazi experimentation, which often involved research on children are well-known. As a reaction to these aberrations, the Nuremberg Code (1947) laid down a series of research principles that stated that persons involved in medical research should have legal capacity to consent, thus effectively ruling out research on children. However, a complete ban on the use of children in research seems unacceptable as well.

In this context the term 'therapeutic orphans' was framed: not allowing children to participate in research because of a strict adherence to the informed consent doctrine would lead to children missing out on certain therapies. Such therapies might only be tested on adults, or children might be prescribed drugs that were untested on them (Avard et al. 2009, Kodish 2005, Pinxten, Nys and Dierickx 2008). The Declaration of Helsinki, which was issued by the World Medical Association in 1964 and subsequently revised, therefore allows research on children, and formulates the conditions under which such research is permitted. In its current revision (2004) the Declaration states that:

> when a subject deemed legally incompetent, such as a minor child, is able to give assent to decisions about participation in research, the investigator must obtain that assent in addition to the consent of the legally authorized representative.

and:

> research on individuals from whom it is not possible to obtain consent, including proxy or advance consent, should be done only if the physical/mental condition that prevents obtaining informed consent is a necessary characteristic of the research population.

Hence, the Declaration allows for specific types of research to be performed using vulnerable subjects such as minors. Similar principles are described in the CIOMS guidelines on biomedical research (Council for International Organizations of Medical Science (CIOMS) 2002):

> before undertaking research involving children, the investigator must ensure that: the research might not equally well be carried out with adults; the purpose of the research is to obtain knowledge relevant to the health needs of children; a parent or legal representative of each child has given permission; the agreement (assent) of each child has been obtained to the extent of the child's capabilities; and, a child's refusal to participate or continue in the research will be respected.

Most ethical literature on children in medical research seems to acknowledge that at least some non-therapeutic research that involves children's participation is allowed, but that they are subjects in need of extra protection. Brown uses the principle of 'future independence': parents can consent to research on their children if the future independence of these children is not impaired (Brown 1982). Other authors refer to the concept of 'minimal harm' or 'best interest' to set some standard on the type of non-therapeutic research that can be performed with children (Kodish 2005, Koocher and Keith-Spiegel 1994, Macklin 1982).

Children and Biobanks

Most of the above mentioned literature and guidelines in the context of medical research were written with clinical trials in mind. Much less is written about the use of pediatric stored tissue for non-therapeutic research. Storage of such samples in a biobank raises a number of ethical questions that are not addressed by these documents. Does the storage and use of genetic material from children, in whatever form, pose specific ethical issues? Is such material needed, or could the problem be avoided altogether by using only material from adults?

Many existing biobanks store and use material from adults only, thus avoiding such issues. Some have argued that, as DNA is stable throughout a lifetime there is no need for pediatric biobanks (Baumann 2001). However, genes are often studied in combination with environmental and phenotypical factors. The influence of prenatal and early childhood environmental factors on gene expression later in life is now generally acknowledged. Moreover, conditions such as allergy, asthma, food intolerance, diabetes and obesity, which occur early on in childhood, are attributed to a combination of genetic and environmental factors. As the latter conditions have nowadays attained epidemiological proportions, systematic and large-scale research on genes and environment, starting in early childhood, can be useful.

Examples of such child-cohort studies are the Avon Longitudinal Study of Parents and Children ('Children of the Nineties'), situated in Bristol, which has collected detailed phenotypic and environmental information from 14,000 babies born in 1991-1992 (Pembrey 2004). Similar studies are 'Generation R' in Rotterdam, involving 10,000 babies, and 'Born in Bradford' which will follow 10,000 Bradford children over the next 20 years. Also, conditions that occur in childhood, such as childhood cancer and specific birth defects are most easily studied using material from the children affected. It would stall research and in many cases be unfeasible to wait for the child to reach the age of majority. An example of disease-specific use of pediatric tissue is the research into genetic factors of birth defects (Jenkins et al. 2008). A special case is the existence of collections of blood samples taken from newborns on blood spot cards. Newborn screening is done on most of babies in many countries: these collections, which are originally gathered for diagnostic purposes, form an easily accessible resource to study the prevalence of certain genes in a population. Klotz has described how these samples could be used to study cancer susceptibility genes (Klotz et al. 2006).

While there is a growing acknowledgement of the importance of storing pediatric samples in biobanks for research uses, there is still considerable debate and uncertainty about how to deal with the ethical issues that are raised by this endeavor. One of the most controversial areas is consent for inclusion and use of samples within a biobank. This is hotly debated area of bioethics, not just for children but for adults as well. However, children raise specific ethical issues, such as a child's capacity to give consent; the scope of the consent; and the type of harms and benefits that may exist for children and yet are different for adults. We will discuss these in detail and make suggestions for the development of good practice in this area.

Consent

Most literature on biobank storage and research agrees that at least some form of informed consent is needed. Although many authors claim that re-contacting donors to seek consent for specific research uses may not be feasible, it is deemed good practice that at least broad consent for future research uses is sought at the time of collection of the tissue. Consent requirements may be less stringent if anonymous samples are used from existing collections (Brock 1994), but even in that case, it is deemed good practice to obtain consent. For pediatric stored tissue, however, it is unclear what this consent might entail. Children are, except perhaps in the case of emancipated or mature minors, incapable of giving legal consent to research.

Who should consent?

An initial question that arises is who should give consent on behalf of the children? Can parents be permitted to give consent to research on their children? Brock sums up some reasons why parents are the appropriate decision makers for research in general (Brock 1994): First of all, children cannot do it, as they have no legal competence. Parents are the ones who will bear the consequences of their decisions regarding their children. Furthermore, parents have the right to raise their children according to their own standards. In most cases, the child would want a parent to decide. Ross says that parents have an interest in making decisions for their child, which involves respect for the child and nurturing their development so that the child can become the kind of person of which the parents approve (Ross 2006).

This is reflected in the ethical literature on children and biobanks, which acknowledges that biobank storage and research can be done under a regime of parental consent (Burke and Diekema 2006, Helgesson 2005, Holm 2005, Knoppers et al. 2002). In addition, guidelines specific to biobanks often mention parents as decision makers. For example, the Nuffield Council of Bioethics (Nuffield Council on Bioethics 1995) refers to 'someone with parental responsibility'. It is unclear whether this means that both parents should give their consent. The familial nature of genetic research might suggest that this is the preferred route. However, this could also mean that the decision should be a familial one, involving siblings as well. But in present-day societies, many children do not live with both genetic parents, and the relationship with one such parent may be strained or nonexistent. So we think that consent from both parents, and even familial consent, could be seen as a best practice rather than a formal requirement.

Scope of consent

If it is acceptable for parents to give permission on behalf of their children, what should be the scope of this consent, both in time and in content? It seems that such consent would only be valid until the child reaches the age of majority. In longitudinal cohort studies, where contact between children and researchers are frequent, it is easy to renew consent at a specific point, but if the research is on genetic information that has been acquired as a one-off event, such as blood spot cards, this is more problematic. Still it seems unfair that researchers can continue to use the genetic data of an 18-year-old because his or her parents gave consent when he or she was an infant. Many authors, such as Burke and Diekema (2006), Helgesson (2005) and Fisher (2006) agree that the proxy consent given by parents is only valid until the child is able to make his or her own decisions. It may in some cases be impractical or impossible to locate the donor. However, we believe that at the time of conception of the pediatric biobank, the process for obtaining re-consent should be laid down as well as the procedure to be followed where the donor can no longer be located and such consent can not be obtained. It may be fair that in that case the material and data is no longer used for research purposes[1]. In addition, the donors whose tissue was included following consent by their parents should be given the opportunity to withdraw their consent if they so wish. As their tissue is being used on the basis of a consent given by someone else, this right to withdraw is more important for children than for adult donors (Holm 2005).

Another aspect of the scope of consent is the question of the nature of the research to which parents should be allowed to consent. Should they be allowed to give consent to general genetic research on tissue from their children, or should this be disease-specific or gene-specific for the child or be restricted to the diseases from which the child actually suffers? In the literature on biobanks in general, some argue that broad consent is acceptable: adult donors should be able to give consent for any future research use (Hansson et al. 2006, Wendler 2006). We doubt whether this is applicable in the case of parental consent. It may seem unfair for parents to give consent to any genetic research on their children's DNA, as DNA is stable throughout a lifetime. Some adults may have strong opinions on the type of research done on their genetic material and at least would want to have some control over that. Moreover, when someone finds out, upon reaching the age of majority, that he or she has been subject to full genetic screening and research, this may be detrimental to that person's trust in both their parents and in biomedical research in general.

---

1 An intermediate solution in this case may be to continue using the data and samples, in an anonymized way. It has been pointed out by some authors, however that anonymization is not completely unproblematic, as subjects may feel strongly about the type of research that is done on their tissue, regardless of the level of identifiability. See, for example, Trouet (2003).

On the other hand, having too much detail about each possible research project that might involve research on their children's DNA may be an unnecessary constraint and overly restrictive. This is particularly true if it involved implementing an extensive 'tick-box'-style consent, and a policy of re-contacting parents for each minor change in research protocol. We believe that a good middle ground is to ask parents to give permission for research on specific conditions or diseases from which their child suffers or may suffer in the future. For example, a parent might give permission for research into the genetic aspects of obesity, but a researcher may not use the same genetic data for research on the 'smoking gene' as well, without re-contacting and informing either the parent or the mature child. The difficulty with this as there are some genetic variants that are responsible for very different conditions, such as ENPP1 that is responsible for obesity and type 2 diabetes (Meyre et al. 2005). Ruling out a specific disease as a subject for research, may inadvertently rule out a valuable line of research enquiry into the way that different genetic variations may be responsible for different conditions. We think that if such a finding occurs, researchers should try to inform children and their parents if this is feasible. If they cannot be contacted, an ethics committee can decide whether the research on the disease for which no specific consent exists, may continue.

Assent

Should minors that are not yet legally competent to give consent, be informed and asked to give their assent for storage and research in biobanks? To what extent should they be given relevant information and should their permission (assent) and refusal (dissent) be taken into account? From what age should this be done? Should this be done based on a fixed age or on a criterion for maturity? And if a fixed starting point is adopted, would it follow that children below that point can only contribute with parental consent, or would it mean that they should be excluded from biobank research altogether? CIOMS, in its general guidelines on research (Council for International Organizations of Medical Science (CIOMS) 2002), suggests that the latter situation should prevail in that case. It states that with very small children it is often difficult to assess the sincerity of objections and they advise limiting research as far as possible to older children. Priscilla Alderson, however, has shown that already very small children are capable of formulating opinions on things, and should therefore be consulted (Alderson 2008).

This raises the question whether a child's dissent could be overridden by his or her parents, in the interests of providing a moral lesson to the child? An oft-quoted example is the one given by Willard Gaylin (Gaylin 1982), who describes a father of a 10-year-old boy who gives permission, in the face of his child's refusal, for blood be taken from his son for research. The father justified his decision thus; 'I'll be damned if I was going to allow my child, because of some idiotic concept of children's rights, to assume that he was entitled to be a selfish, narcissistic little bastard'. This father may have had a point, but consider the following scenario:

What if this research would interfere with other activities – perhaps seemingly trivial to adult eyes – that are very important to a child, such as a birthday party? In this case, we would advise listening to the child instead. And what of the case of a three-year old who has a phobia about needles? It is in this case it is probably advisable not to enroll the child in research that requires taking a blood sample.

Given the fact that biobank research is non-therapeutic and does not entail direct benefit for the child in question, we would suggest that it is prudent to be very careful in such situations. Moreover, the United Nations Convention on the Rights of the Child (1998) asserts that children have a right to say what they think should happen when adults make decisions that affect them (Article 12). They also have the right to obtain and share information (Article 13). As the convention is legally binding in most countries worldwide, biobank policies should also include policies addressing how children should be consulted and informed, according to their different age groups and levels of maturity. Problems may arise where the subject group shares a specific characteristic such as being uncomfortable with or antithetical to new or unfamiliar procedures. For example, having blood taken may cause more anxiety to autistic children who often experience high levels of discomfort in unfamiliar situations than it would to children who don't suffer from that condition. Special care should be taken to make such research as comfortable as possible for the child as possible for example, by taking samples in the child's home environment; using mouth swabs rather than an injection to draw blood; and using someone known to the child, such as the family physician.

Minimal Harm and Benefit

A principle that is often quoted in the literature on non-therapeutic research with children is that of 'minimal harm'. Only research that poses no more than minimal harm to the children involved is allowed. But there is no agreed position in the literature as to what constitutes minimal harm for children involved in biobank research. For example, empirical data from Neidich (Neidich et al. 2008) seems to suggest that many people consider research on genetic information in a biobank as minimal harm. On the other hand, authors such as Baumann (2001) adhere to some form of genetic exceptionalism; genetic information is special, contains the entire future of an individual, and can be abused by insurers and employers. Hence, the risk encountered by biobank research is higher than minimal risk and children should not be enrolled.

Emotional and physical harm

There are two categories of possible risks that can be identified. Firstly, there is the risk of emotional and physical harm, and secondly there is the risk of breaches of confidentiality. As mentioned above, children may experience fear when they are placed in an unfamiliar environment. Moreover, children whose days are already taken up with a full schedule of activities may be overburdened by participation in research, such as longitudinal cohort studies, that requires frequent contact with researchers (Kaufman et al. 2008). Where the biobank research involves the use of questionnaires to gather phenotypical or other kinds of information, there is a risk that some questions may make children feel uncomfortable. What is children consider private and sensitive information may differ from what their parents or the researchers think, as pointed out by Williamson (Williamson et al. 2004). We think physical and emotional harm can be minimized if researchers consider carefully the sensitivities of different age groups. Research should be as minimally invasive as possible and ideally conducted in a familiar environment.

Confidentiality

The widely acknowledged foremost risk with regard to biobank research is related to confidentiality. As genetic information contains clues to possible future diseases and conditions, it is thought to be eagerly sought by insurers, employers and the government (O'Neill 2002). Hence, most documents stress the need for the security of this information, and the disclosure to donors of the privacy protections in place. Such protections may involve irreversible anonymization of the data, but this is only applicable to certain studies, such as one- off, single -hypothesis research projects. Biobanks involving children's genetic material will mostly be either disease-specific or part of longitudinal cohort studies, which involve on-going monitoring of the child or collection of data from a number of different sources. In this case, anonymization is impossible as there is frequent contact between the children and the researchers. In on-off projects, data could be coded and the access to identifiable information could be restricted to specific researchers. However, some way of identifying the donors and linking with medical data will often be necessary.

With regard to genetic information, Roche contends that we should be concerned that others know more about ourselves than us, and calls for strict privacy protections (Roche and Annas 2006). But Manson and O'Neill (2007) have argued that the privacy risks related to genetic information are no more nor less than those encountered in providing other health information. They argue that by looking at a specific person alone, some information about his or her (future) health can be deduced. In this respect, biobank research would not pose greater risks than those encountered in everyday life, by children and adults alike.

The other risk that is often mentioned with regard to biobank research is that of group stigmatization (Burke and Diekema 2006, Holm 2005). This would affect children of the stigmatized group anyway, regardless of whether they personally contributed to the research or not. We do not think that either the issue of confidentiality or group stigmatization is an insurmountable reason not to include children in biobank research. These risks should be properly addressed through general biobank policies and by facilitating rigorous public debate on the research.

Benefit

One of the general principles that is often considered with regard to research on children is that there should be some direct benefit to the participants. Brock describes three potential benefits to children when they participate in research. These are; direct medical benefits to the child itself; non-medical benefits to the child, for example, the child's altruistic act may help his or her social and moral development; and finally benefits may flow to others. Some argue that there is an ethical obligation for children to participate in research, because they themselves benefit from research conducted on previous generations of children (Brock 1994). In the case of biobanks, there is little direct benefit to be gained for the participants, although indirect benefits may accrue, as is the case in longitudinal cohort studies. Here the children may benefit from the regular health checkups (Kaufman et al. 2008). The second benefit that of social and moral development, is less easy to appreciate. Where, for example, material is used that was originally gathered for other purposes, or if no further contact with the child is attempted, the impact on his or her socialization will be minimal. However, if appropriate information is given to the child as to how their donation will be used, and proper assent is sought, this can increase the sense of self-worth of the child and his or her (future) trust in biomedical research. The third principle is not concerned with direct benefit to the child. It corresponds with some of the general principles that are quoted by CIOMS (Council for International Organizations of Medical Science (CIOMS) 2002) and the Declaration of Helsinki (World Medical Association (WMA) 1964): that research on children should benefit children of the same age or with the same condition.

As we have argued in the section on research, we think this principle is applicable to biobanks. It makes sense, in our opinion, to allow parents to consent to genetic research on diseases and conditions that are well-specified at the time of consent, but not to allow researchers to use the pediatric genetic materials for all kinds of genetic research. However, this criterion must not be so strict as to disallow longitudinal cohort studies. In these types of studies, the conditions that are being researched can be well-defined. In addition, establishing renewed contact with parents or children is relatively unproblematic if new conditions are to be studied. Therefore it is essential that good governance structures are put in place to allow this to happen.

The issues that policy documents should address at the time of inception of a pediatric biobank include an explanation of the need for and a description of the type of information that will be given to children of different age groups. As well as being fully informed, the child should be permitted to express his or her feelings with regard to the collection of tissue and the subsequent research, when feasible. This should be accompanied by parental consent, preferably by both parents. Parents should be allowed to consent to research on their children's DNA and tissue in relation to specific diseases, genes and conditions. They should not be asked to give 'broad' consent to any possible future research on their children's genes. If researchers want to use pediatric tissue for other purposes than those originally proposed, they should ideally re-contact the parents and their children. In any case, youngsters should be re-contacted when they are 18-years-old to give consent on the use of their tissue. If this is not possible, an ethics committee should decide on the acceptability of the study. We believe that the principle of minimal harm should also apply to the use of tissue from children in research. Preferably, tissue that is taken in a medical context should be used. If this is not feasible, buccal swaps can be used rather than taking extra blood. If possible, medical examinations for research could be done in the home environment by a physician who is known to the child rather than an unfamiliar researcher.

Conclusion

The storage of tissues from minors in biobanks and the subsequent research, raise ethical issues that do not arise in discussions on the use of tissue donated by adults. Specifically, questions of consent, the requirement in pediatric research of posing no more than minimal harm, and providing benefits to either the child or the group to which he or she belongs need further scrutiny. However, if proper policies and governance structures are put in place to address these issues, and if the children are approached in a way that is appropriate to their age and maturity, there is no need for a ban on using their genetic material for certain types of research. On the contrary, such research may provide answers to important questions about diseases and conditions that affect children worldwide.

Acknowledgements

This work was supported by GeneBanC, an EU-FP6 supported STREP contract number 036751 and FWO Flanders, project number G029107.

Bibliography

Alderson, P. 2008. *Young Children's Rights. Exploring Beliefs, Principles and Practice*. London/Philadelphia: Jessica Kingsley Publishers.

Avard, D., Silverstein, T., Sillon, G. and Joly, Y. 2009. Researchers' perceptions of the ethical implications of pharmacogenomics research with children. *Public Health Genomics*, 191-201.

Baumann, T.K. 2001. Proxy consent and a national DNA databank: An unethical and discriminatory combination. *Iowa Law Review*, 86(2), 667-701.

Brock, D.W. 1994. Ethical issues in exposing to risks in research, in *Children as Research Subjects. Science, Ethics & Law*, edited by M.A. Grodin and L.H. Glantz. New York/Oxford: Oxford University Press, 81-101.

Brown, P.G. 1982. Human independence and parental proxy consent, in *Who Speaks for the Child? The Problems of Proxy Consent*, edited by W. Gaylin and R. Macklin. New York/London: Plenum Press, 209-222.

Burke, W. and Diekema, D.S. 2006. Ethical issues arising from the participation of children in genetic research. *Journal of Pediatrics*, 149(1Suppl), 34-38.

Council for International Organizations of Medical Science (CIOMS). 2002. *International Ethical Guidelines for Biomedical Research Involving Human Subjects*. Geneva.

Fisher, C.B. 2006. Privacy and ethics in pediatric environmental health research – Part II: Protecting families and communities. *Environmental Health Perspectives*, 114(10), 1622-1625.

Gaylin, W. 1982. Competence – No longer all or none, in *Who Speaks for the Child? The Problems of Proxy Consent*, edited by W. Gaylin and R. Macklin. New York/London: Plenum Press, 27-54.

Hansson, M.G., Dillner, J., Bartram, C.R., Carlson, J.A. and Helgesson, G. 2006. Should donors be allowed to give broad consent to future biobank research? *Lancet Oncology*, 7, 266-269.

Helgesson, G. 2005. Children, longitudinal studies, and informed consent. *Medicine, Health Care and Philosophy*, 8(3), 307.

Holm, S. 2005. Informed consent and the bio-banking of material from children. *Genomics, Society and Policy*, 1(1), 16-26.

Jenkins, M.M., Rasmussen, S.A., Moore, C.A. and Honein, M.A. 2008. Ethical issues raised by incorporation of genetics into the National Birth Defects Prevention Study. *American Journal of Medical Genetics, Part C Seminars in Medical Genetics*, 148C, 40-46.

Kaufman, D., Geller, G., LeRoy, L., Murphy, J., Scott, J. and Hudson, K. 2008. Ethical implications of including children in a large biobank for genetic-epidemiologic research: A qualitative study of public opinion. *American Journal of Medical Genetics, Part C Seminars in Medical Genetics*, 148C, 31-39.

Klotz, J., Bryant, P., Wilcox, H.B., Dillon, M., Wolf, B. and Fagliano, J. 2006. Population-based retrieval of newborn dried blood spots for researching paediatric cancer susceptibility genes. *Pediatric and Perinatal Epidemiology*, 20(5), 449-452.

Knoppers, B.M., Avard, D., Cardinal, G. and Glass, K. C. 2002. Science and society: Children and incompetent adults in genetic research: Consent and safeguards. *Nature Reviews Genetics*, 3(3), 221-225.

Kodish, E. 2005. Ethics and research with children: An introduction, in *Ethics and research with children. A case-based approach*, edited by E. Kodish. Oxford/New York: Oxford University Press, 3-25.

Koocher, G.P. and Keith-Spiegel, P. 1994. Scientific issues in psychosocial and educational research with children, in *Children as Research Subjects. Science, Ethics & Law*, edited by M.A. Grodin and L.H. Glantz. New York/Oxford: Oxford University Press, 47-80.

Lederer, S.E. and Grodin, M.A. 1994. Historical overview: Pediatric experimentation, in *Children as Research Subjects. Science, Ethics & Law*, edited by M.A. Grodin and L.H. Glantz. New York/Oxford: Oxford University Press, 3-25.

Macklin, R. 1982. Return to the best interests of the child, in *Who Speaks for the Child? The Problems of Proxy Consent*, edited by W. Gaylin and R. Macklin. New York/London: Plenum Press.

Manson, N.C. and O'Neill, O. 2007. *Rethinking Informed Consent in Bioethics*. Cambridge: Cambridge University Press.

Meyre, D., Bouatia-Naji, N. and Tounian, A., et al. 2005. Variants of ENPP1 are associated with childhood and adult obesity and increase the risk of glucose intolerance and type 2 diabetes. *Nat Genet*, 37(8), 863-867.

Neidich, A.B., Joseph, J.W., Ober, C. and Ross, L.F. 2008. Empirical data about women's attitudes towards a hypothetical pediatric biobank. *American Journal of Medical Genetics Part A*, 146(3), 297-304.

Nuffield Council on Bioethics. 1995. *Human Tissue, Ethical and Legal Issues*.

O'Neill, O. 2002. *Autonomy and trust in bioethics*. Cambridge: Cambridge University Press.

Pembrey, M. 2004. The Avon Longitudinal Study of Parents and Children (ALSPAC): A resource for genetic epidemiology. *European Journal of Endocrinology*, 151(Suppl_3), 125-129.

Pinxten, W., Nys, H. and Dierickx, K. 2008. Regulating trust in pediatric clinical trial. *Medicine, Health Care and Philosophy*, 11(4), 439-444.

Roche, P. A. and Annas, G. J. 2006. DNA testing, banking, and genetic privacy. *New England Journal of Medicine*, 355(6), 545-546.

Ross, L.F. 2006. *Children in medical research*. Oxford/New York: Oxford University Press.

The Nuremberg Code (1947).

United Nations Convention on the Rights of the Child (1998).

Wendler, D. 2006. One-time general consent for research on biological samples. *British Medical Journal*, 332(7540), 544-547.

Williamson, E., Goodenough, T., Kent, J. and Ashcroft, R. 2004. Children's participation in genetic epidemiology, in *Genetic databases: Socio-ethical issues in the collection and use of DNA*, edited by R. Tutton and O. Corrigan. London/New York: Routledge, 139-160.

World Medical Association (WMA). 1964. *Declaration of Helsinki. Ethical Principles for Medical Research Involving Human Subjects*.

Chapter 8
Deciding Whether to Participate in a Biobank: The Concerns of Healthy Volunteers

R. Jean Cadigan and Arlene M. Davis

This chapter adds to a growing body of qualitative empirical literature that investigates individuals' motivations to participate in biobanks and emphasizes the social context of individual decision-making regarding participation (Barr 2006, Haimes and Whong-Barr 2004, Whong-Barr and Haimes 2003, Hoeyer 2004, 2003). However, unlike many of these studies, this chapter presents findings from a study of healthy volunteers recruited outside of a clinic setting, who were approached to enroll in the Environmental Polymorphisms Registry (EPR), based in North Carolina in the US. Two key issues that emerged from this study have implications for the design of recruitment and enrollment protocols involving healthy individuals for a biobank of this kind. The first issue was the role of financial incentives in affecting individuals' decision-making whether to donate a sample of blood to the biobank. The second was how the consideration of external factors, such as the governance mechanisms in place for medical research, was brought into play when individuals were in the process of deliberation. These findings are significant as they highlight the reasons why healthy volunteers decide to participate in a biobank and the reasons why they decline participation. As recruitment of this population into biobanks as controls or as part of a sample population increases, healthy volunteers' perceptions of participation in light of the ethical issues biobanking raises are especially important to understand.

The Environmental Polymorphisms Registry

The Environmental Polymorphisms Registry was created by the NIEHS in collaboration with the General Clinical Research Center (GCRC) at the University of North Carolina at Chapel Hill. The goal of the EPR is to assemble a demographically diverse collection of DNA samples from 20,000 adults in North Carolina, both patients and healthy volunteers, with associated data on their race, ethnicity, gender, and age (Chulada et al. 2008, US National Institute of Environmental Health Sciences 2006).

Participants' personal information is linked to their samples in coded form and stored up to 25 years as NIEHS property. Samples are primarily available to researchers interested in environmental response genes.[1] Upon application, a researcher receives samples without identifiers for genotyping. When genotypes of interest are ascertained, the researcher's follow-up study protocol is first reviewed by an EPR Steering Committee for scientific merit, scientific design, and possible subject participation. (EPR subjects are not asked to participate in more than one study at a time). Upon favorable review, the protocol proceeds to relevant institutional and research ethics board reviews. When all approvals are obtained, the researcher receives the EPR subject contact information necessary for recruitment into that particular study. The EPR informed consent form makes clear that participation in such future studies requires a separate consent process.

The EPR recruits healthy volunteers several ways, including via mass emails to University of North Carolina faculty, staff, and students. Individuals are encouraged to contact EPR representatives with questions and to schedule an enrollment appointment at the GCRC. The appointment involves a formal discussion about the consent process, data collection with an EPR representative, and finally the taking of a blood sample. Those who join receive 20 dollars (US$20) in cash.

The EPR offers two participant safeguards that are part of the oversight framework applied to all medical research in the US. The first is approval by a research ethics committee, called an Institutional Review Board (IRB). IRBs were established to ensure compliance with ethical principles and regulatory requirements for human subject research. They aim to safeguard the rights and welfare of individuals in several ways, including review of informed consent processes (US Department of Health and Human Services 2005). Informed consent and IRB approval have become the 'twin procedural mechanisms' to ensure the protection of research individuals (Boulton and Parker 2007: 2190).

The EPR has also obtained a second subject safeguard available in the US, a Certificate of Confidentiality issued by the National Institutes of Health (NIH). Certificates aim to protect identifiable, sensitive research information from forced disclosure in any civil, criminal, administrative, legislative, or other proceeding, whether at the federal, state, or local level (US Department of Health and Human Services 2003); they have been recommended for all biobanks to protect their participants (US National Cancer Institute 2007). The Certificate permits researchers to refuse to disclose identifying information that could have adverse consequences for individuals or damage their financial standing, employability, insurability, or reputation. Certificates of Confidentiality are intended to help achieve research objectives and promote participation in sensitive studies by assuring confidentiality and privacy to participants. While rarely tested, a recent North Carolina court case offers reason to be wary about their full legal protection (Beskow, Dame and Costello 2008).

1 The environmental response genes are those genes that interact with environmental factors to cause complex disease such as asthma, cancer, or diabetes.

The Interview Study

Our study included both semi-structured interviews and a self-administered questionnaire with 29 individuals who considered enrollment in the EPR. We also conducted observations of the EPR enrollment process in the GCRC.[2] We sought to compare the perspectives of individuals who enrolled in the biobank with those who declined participation, given that the latter group is often overlooked in the ethics literature.

We recruited individuals by sending a mass email to the same University faculty, staff, and students who had recently received an EPR recruitment email. When interested individuals contacted us, we assessed eligibility based upon current EPR enrollment or a decision to decline participation. For individuals who had not joined the EPR, further screening excluded those who chose not to participate due only to inconvenience or other logistic burdens. Surprisingly, we were contacted by individuals who either did not *know* they had joined the EPR until we provided some details of the enrollment process, or who knew only that they 'gave blood somewhere'. While these individuals were ineligible for our study, their existence points to the possibility that some people found EPR enrollment so inconsequential that it had been largely forgotten.

Our self-administered questionnaire gathered basic demographic information and assessed attitudes toward genetic research. The interviews consisted of approximately 60 questions; with few exceptions, those who joined the EPR and those who did not were asked the same questions. Questions were grouped in several categories: reasons for accepting or declining participation; perceptions of personal benefits and risks as well as benefits or risks for families, special groups, or society; and understanding of key features of the EPR. Interviews were conducted at a location chosen by the participant, lasted from 45 to 75 minutes, and were transcribed verbatim.[3] All individuals received US$35 for participating.

The 29 participants included 15 people who joined the EPR ('joiners') and 14 people who chose not to join ('decliners'). Joiners and decliners were very similar demographically in age, level of education, and whether they were fulltime employees or students.[4] The majority of both groups had experience working in

---

2 Both authors are associated with The Center for Genomics and Society at the University of North Carolina, Chapel Hill. Neither author has been associated with the EPR. The UNC IRB approved our interview study.

3 The authors (Cadigan and Davis) conducted the majority of the interviews; two colleagues associated with the study conducted the rest.

4 The age range of joiners was 22-58, with an average of 36; decliners' range was 21-59, with an average of 32. Seventeen participants were employed full-time (10 joiners and seven decliners), and 10 were full-time students (four joiners and six decliners). Twenty-one subjects had at least a college degree (10 joiners and 11 decliners). Participants included 19 females (eight joiners and 11 decliners) and 10 males (seven joiners and three decliners).

the healthcare industry,[5] as well as substantial and positive previous experience participating in research studies.[6] Finally, questionnaire responses indicated that their attitudes toward genetic research were similar. Both groups viewed participation in genetic research as important and reported valuing the research institution's reputation when considering participation. While both groups viewed genetic research as useful for cure and treatment of disease, they also saw it as potentially harmful to people.

Thus, our participants are a select group of individuals, drawn from the equally select EPR recruitment population, who reflect membership in a research university. In fact, this population receives recruitment emails, like ours, almost daily for University-based studies. These IRB-approved emails all include brief information on purpose, eligibility, procedures, and payment.

Findings

Our study revealed striking similarities between the perspectives of individuals who enrolled in the EPR and those who declined participation, including their identification of risks and benefits that related to participation. They were similarly knowledgeable of key features of the biobank, suggesting that the informed consent process imparts little additional knowledge to those who join. Our findings also indicate that people who joined the biobank were motivated by payment and often framed risk and benefit in the short-term. Payment for participation, combined with ease of participation, was regarded as beneficial, and any concerns over data security were minimized by trusting the safeguards associated with the study. In contrast, those who declined participation, while often able to identify short-term benefits, did not view current safeguards as sufficient to protect their DNA sample, and genetic and personal information, in the long-term. Our findings have significance for the design of recruitment and enrollment procedures, in particular understanding how individuals obtain information to aid their decision-making; the importance for individuals of knowing the procedures and oversight mechanisms in place; and the motivations that underpin their decisions to participate. We will discuss each of these in turn below.

5 Most of the students in our study were graduate students who had prior work experience, including work in the healthcare industry.

6 Fourteen joiners and 12 decliners had previous research experience, including two joiners and three decliners who had participated in genetic studies. Of these 26 people, 22 rated their previous research experience as good.

When does Decision-making Occur?

While the formal consent process conducted with an EPR representative permits potential individuals to obtain and weigh information for decision-making, our data indicate that decision-making occurred well before this event. The majority of both joiners and decliners relied upon their own background research to make their decision. Nine of the 14 decliners were initially interested in joining, but sought more information. Some called EPR representatives while others visited the study's website. Those who declined typically did so because they found information that led them to doubt the security of their genetic information. One decliner, who called an EPR representative, stated:

> Basically I wanted to know essentially what the identification protocol was, and how the records were retained, and who had access to what. And, the final question was ultimately the question that made up my mind: Will they ever be able to identify me with this information?

Another decliner commented:

> I pursued the information because it was very difficult for me to decide not to join. Because, I really believe that it's a very valuable collection of data … But, I don't want [to be put] in this kind of situation where there's the association [between my name and sample] and there's a potential for abuse.

Nine of the 15 joiners visited the EPR website or called a representative for more information. Some sought a particular type of information, such as study purpose, while others read through the consent materials in detail. Another woman chose to participate after getting details from a co-worker who had enrolled. Thus, the majority of those in our study gathered the amount and type of information they viewed as relevant to decision-making outside the setting of a personal meeting with a study representative, the context often considered the best suited to considerations of study participation (Coleman et al. 2005: 345).

The importance of information gathering and decision-making for individuals before an enrollment appointment is well represented in the literature. In addition, and consistent with Hoeyer's (2003: 231) observation that few donors to a population-based biobank in northern Sweden read the consent form, our GCRC observations demonstrated that some participants were ready to sign a consent form, the physical evidence of agreement to participate, before reading it. However, the EPR enrollment procedure did not deviate when participants were eager to sign; each time the study recruiter reviewed the consent information before participants were permitted to sign.

Whether people were ready to sign the consent form before this review because they had read it beforehand or because its content was unimportant to them is unclear from our observations. However, the attempt to sign the form before review suggests that decision-making had already occurred. Further, two joiners did not change their decision to participate even when confronted, during the consent process, with information contrary to their perceptions of the study. Both were unaware that the EPR is a longitudinal study. For example, one commented:

> I was surprised though, I didn't realize that it was a call back study or that they were going to maintain some information. I thought maybe they would ask me what zip code I grew up in, something like that – sort of more general demographic information rather than keeping all of my contact information … I thought that if they were going to do something like that, it should have been included in the advertisement, because I think the advertisement sort of conveyed that this was a one-time thing. You go in. You give some blood. Maybe fill out a survey. But, it wasn't something that they would contact you about later. That would be the end of it.

It may seem surprising that finding the advertisement misleading regarding an important point did not deter enrollment. However, we suggest that the threshold for declining participation at an enrollment appointment is quite high given the time and energy an individual has invested at that point, as well as the potential embarrassment of rejecting a study in front of the study's representative.

Finally, regardless of the EPR's thorough consent procedure, we found little difference between joiners and decliners in their understanding of over 10 key features of the EPR. These features, including duration of sample storage, sample ownership, researcher access to samples, study withdrawal, and disclosure of DNA analysis, were identified by reviewing the EPR consent form in light of research ethics literature. When asked, both joiners and decliners knew little about the features. Joiners seldom knew more than decliners, suggesting that the formal consent process may impart no more knowledge about most of the features than what individuals gain on their own. However, two exceptions are important: joiners were far more likely to know that participants would not receive results from their initial DNA analysis, a misattributed personal benefit (Churchill et al. 2003), and were slightly more likely than decliners to know that participants could withdraw from the EPR, a concept central to O'Neill's (2003) concerns regarding deception and coercion that we will discuss further below.

Motivators to Join and Consideration of Benefits and Risks

When asked why they chose to participate in the EPR, the vast majority said they joined for the US$20; a few also credited the study's potential benefit to society, the convenience of joining, and their desire to support the University. Despite any benefits they may have associated with participation, decliners focused on concerns regarding confidentiality and privacy. For example, one decliner noted, 'Although I think it's a really good idea to have a polymorphism registry, I don't think at this time there are legal safeguards in place that make me comfortable giving my DNA out. And, I just don't quite trust it'. The majority of both joiners and decliners were also influenced by their background or past experiences. For example, six joiners cited their science background as enabling them to feel comfortable participating, while five decliners said their science background made them troubled by breaches in data security or privacy they had seen or could envision.

All respondents were asked a series of questions regarding potential benefits or risks they considered in their decision. The majority of joiners and decliners thought the EPR will benefit society (although as noted above, only a few cited this as a motivation to join), but few considered personal, familial, or group benefits. Some joiners regarded a potential to learn about their genetic information as a personal benefit, although the EPR does not promise that any studies would include personal results. Several joiners and decliners viewed the US$20 as a personal benefit.[7] For example, when asked whether he thought of any personal benefits when considering participation, one joiner commented, 'Yes, but only because of the money. Not for the research itself'. Another stated, 'The only personal benefit I saw was that maybe I'll have some set of polymorphisms that will make them bring me back for another study where they'll pay me more money'.

Apart from payment issues, individuals in our study raised typical – and very general – ideas about potential benefit. However, both groups detailed very similar lists of potential concerns that mirror those in research ethics literature. While both joiners and decliners cited the same concerns, they were decisive for decliners. Recent US studies have found that the general public is both supportive of genetic research because of the promise of new treatments for diseases, while at the same time concerned about related discrimination, government or corporate exploitation of genetic technologies, government access to genetic information, researchers 'playing God', and the privacy of genetic information (Bates et al. 2005, Gottweis 2002, Tambor et al. 2002, Anderlik and Rothstein 2001). In our study, privacy concerns engaged both groups. One joiner worried that his DNA 'will be on file somewhere. I know that there's a lot of talk about how there's no way to tie it back, but if the right court orders … the data be tied back to the original person's name, it'll happen'. Respondents' concerns also focused on control over the sample:

> Once they had a DNA sample or a blood sample, I didn't know what was going to happen specifically to that material. … I would be worried that because of convenience it would be put to some use that was not specifically a goal of the study for which it was given. [*decliner*]

Other concerns centered specifically on genetic discrimination. A joiner voiced concern that 'if my DNA was able to help identify some kind of genetic

---

7 Although research ethicists are concerned when payment is characterized as a personal benefit of research participation due to the possibility of it being construed as inducement, it is not surprising that many subjects view payment as beneficial.

predisposition to a disease, that an insurance company or some other entity would then use that information to discriminate'.

The similarity of the lists of benefits and concerns generated by joiners and decliners is striking. In trying to identify what may account for those who weighed them in favor of joining versus those who chose to decline, a kind of calibration of risk and benefit emerges. Many decliners had wanted to join the EPR, and some saw the US$20 as a personal benefit. However, for decliners, neither potential benefit nor payment could overcome their concerns about privacy protection in the context of genetic information, especially because of the longitudinal nature of the EPR. Conversely, joiners spoke less about the EPR being a longitudinal study and focused instead on the present experience: ease of enrollment and the immediate benefit of the payment. For joiners, the payment could often outweigh any risk concerns because it was coupled with a reliance – a level of trust – in external safeguards. That trust was bolstered for those whose past experience with such entities was favorable; it was undermined when past experience was not. Thus, the checkered influence of these protections – IRBs, Certificates of Confidentiality, and institutional reputation – are explored below.

Trust in the IRB, Certificate of Confidentiality, and Research Organization

The knowledge and appreciation of IRB review and Certificates of Confidentiality for those in our study is likely reflective of a subject population experienced in healthcare and research participation. Several joiners spontaneously talked about IRB review and the Certificate of Confidentiality as reasons to participate; neither was mentioned in our questions. One joiner, in explaining factors influential to her decision, described the study's convenience then stated, 'I know how protective IRBs are, so I knew there's probably very little risk'. She later added, 'I think if I didn't have a lot of experience working with IRBs and I didn't know how protective they were of people's information and their samples, I might be a little more worried about my genetic information being used to my detriment in the future'. Other joiners felt confident participating because of the University's involvement, mistakenly believing that the University governed the EPR rather than the NIEHS. This misperception is unfortunate given that several individuals indicated their wariness of any government involvement in genetic sample collection.[8]

Two joiners placed trust in federal legislation, stating that HIPAA (the Health Insurance Portability and Accountability Act of 1996) (US Department of Health and Human Services 2008) was protective of their genetic information. One commented on her lack of concern that researchers examining her DNA would gain access to her identifying information: 'Because of HIPAA laws my

8 Subjects also distinguished between academic and commercial banks in terms of trust, expressing weariness with companies potentially profiting from individuals' DNA samples.

personal information wouldn't be shared anyway'. Another remarked that he felt comfortable participating in the EPR because HIPAA will protect 'people who choose to participate in these studies so that researchers can learn more about DNA and how it relates to different diseases'. In fact, HIPAA's application to information collected for research is limited, particularly as data transfer from one researcher to another.

Several decliners, conversely, discussed their doubt regarding the sufficiency of protection under the Certificate of Confidentiality. One noted that the 25 year storage of DNA samples was a primary reason she chose not to participate, and then mentioned her lack of trust in the Certificate of Confidentiality: 'A lot changes in 25 years, and the Certificate of Confidentiality … might be able to be overridden'. Another decliner said she would have joined the EPR had participation been limited to the initial analysis and the DNA destroyed shortly thereafter: 'I think the protection from the Certificate of Confidentiality could not have been challenged and overturned in that length of time. But it's impossible to predict five, six, seven years down the road how our values may change'. She added:

> I think my belief that all laws and protections are contextual made me decide against [participating] … The recent events with the Patriot Act, the Homeland Security, the widespread use of DNA technology in criminal investigations – frankly, right now we may value confidentiality and protect the DNA at the expense of homeland security or a criminal investigation. I think it is entirely likely that in the future that value may change.

Reflecting the social context of decision-making, this decliner and others referenced the current US political climate when discussing their reasoning in declining participation.[9]

Money Matters: The Influence of US$20

One of the most surprising findings in our study was the influence of the US$20 offered to those who participate in the EPR:[10]

> I know it is voluntary. I did it voluntarily. But, I was also motivated by the 20 dollars … Maybe if they took the 20 dollar incentive out, more people would do it … more thoughtfully. I don't know. But, like I said, it was so easy. It was "Oh, come on down. We're doing this. And, this is what it's about. All done. Just a

9 Hoeyer (2003: 238), too, found that Swedish subjects' decision to participate in a biobank was situated in the social and political history of the area.

10 A survey of participants of another US large-scale genetic database found that the payment of US$20 greatly influenced over a third of their subjects' decision to participate (McCarty et al. 2007).

> little bit of blood and you can be on your way". I mean, they did put everything in the [consent] form, but I think you don't have time to process everything.

Twelve joiners said they chose to participate because of the US$20, some even admitting 'If there had been no financial incentive, I wouldn't have done it'. Those who did not cite the money as the reason they joined the EPR still acknowledged its influence in their enrollment decision. Although only one of our questions asked of its importance in the decision to participate, talk of the US$20 permeated the interview. Individuals referred to it when discussing their recall of the EPR recruitment email, the consent process, the ease of decision-making, and when asked to describe the EPR. For example, when stating he had no questions during the consent process, one joiner said, 'It was pretty simple honestly. I've done a number of [research] studies … So, I mean it wasn't just that it was 20 dollars. It was 20 dollars for very little pain and very little time'. Another joiner admitted that he could not remember whether he signed a consent form and that his understanding of the consent process was limited to the understanding of being paid US$20: 'I don't remember, I must have signed some sort of a consent form that I understood. And … I understood that I was going to be paid 20 dollars for doing this'. Later in the interview he commented:

> I've had plenty of blood drawn to do different tests, and I have no idea where it goes. I just get the results of it. And, in this case, the result is a 20 dollar bill that's given to me. And, all my concerns are taken care of and I'm fine.

When asked how they would describe the EPR to someone contemplating joining, several joiners emphasized the US$20. For example, 'Well, I [would] explain how much blood that they take from you. And, the benefits – monetary benefit. And, it's for future generations. There's nothing to be worried about'. Another said his description would 'depend on what part of [the EPR] they were interested in. If they were interested in getting 20 dollars, I'd say, 'It's really easy. You go over [to the clinic], they … kind of walk you through the consent form. And, you just give some blood and you go home'.

The majority of joiners said that their decision to participate was an easy one to make despite their concerns. Some pushed aside concerns because of the influence of the US$20:

> [I had] general low grade fears about my genetic information being on file, … things like, "But, maybe someday I'll be denied insurance …" But, the likelihood that such a thing would ever come to pass is extraordinarily low. And, although they occurred to me, I eventually dismissed them, or actually I think quite quickly dismissed them. And, it was mostly that it's an easy way to make 20 bucks.

Another joiner's reasoning mitigated long-term risk in favor of immediate benefit. He acknowledged his concern that future uses of DNA could be 'potentially damaging, sensitive, and things like that. And then, I was like "Oh well, they pay us". And it's not like they're going to have my name. It's just going to be some number. And then, my name can only be connected by someone who has way more control'.

Conversely, a decliner who wanted to join the EPR because of the payment demonstrated the opposite assessment of long-term risk versus immediate benefit:

> Actually I thought about [joining] for a while because, I don't remember exactly how much it paid, but I seem to remember it paid pretty decently for going and spending ten minutes and then taking a teaspoon of blood. And, I'm a broke college student. I get in as many research studies as I can … At first I was like "Yeah. It's going to be really easy. I'll make like 40 [sic] bucks or whatever to go and do this". And then, I thought about it a while … and, I decided not to because I didn't want my name linked to all of my information like that.

In our examination of decision-making, the influence of payment sharpens the calibration of risk and benefit that allows short-term monetary gain and ease of participation to outweigh concerns regarding long-term use of identifiable DNA samples. This is especially true for those with trust in external safeguards. But some joiners apparently appraised EPR participation as essentially no different from participating in studies without a genetic focus: 'What it comes down to is that they take … a quick blood sample … and they pay you a 20 dollar bill at the end of it'. Of course, participating in a study that links DNA to donors and stores it for 25 years is quite different than, for example, donating blood to a blood bank. Individuals who make this distinction, and who do not trust in external safeguards, find that US$20 does not sway them to participate.

Discussion

Some social scientists have critiqued the biomedical model of consent, arguing that it does not recognize how individuals make decisions about participation relative to their specific needs, wants and experiences (Barr 2006, Corrigan 2003, Dixon-Woods et al. 2007, Miller and Boulton 2007). It has also been argued that this model fails to appreciate whether, and to what extent, individuals base their decision to participate on their trust in the research organization, or other factors that safeguard their interests (Dixon-Woods et al. 2007, Busby 2004: 49-50, Hoeyer 2003, Hoeyer and Lynöe 2006). Our findings add to the qualitative empirical literature on participation in biobanks and other genetic research studies that points to limitations with the current informed consent processes (Barr 2006, Busby 2004, Dixon-Woods et al. 2007, Haimes and Whong-Barr 2004, Hoeyer

2003). Like these studies, our findings demonstrate that individuals often do not read the information contained in consent forms (Barr 2006, Hoeyer 2003), have little understanding of key study features (Barr 2006, Dixon-Woods et al. 2007), and when deciding to participate choose to place trust in the research organization or another entity associated with the study (Busby 2004, Dixon-Woods et al. 2007, Hoeyer 2003).

This brings into question the ethical basis for the informed consent process and whether the requirements for informed consent forms and procedures that have become part of the governance framework for research are effective and whether they actually inform possible participants sufficiently so that they can make a decision that is informed and free of coercion and deception (O'Neill 2003). In the United States, regulations require voluntary participation by individuals who have given consent (or by others acting as proxies or surrogates on their behalf) after receiving study information relevant to their decision to participate (US Department of Health and Human Services 2005). Interpretation of these regulations has created a body of literature debating the kinds and depth of information that potential biobank participants, as opposed to researchers, might consider relevant for individual decision-making (Caulfield 2007, Hansson et al. 2006, Wendler 2006).

Debates regarding payment for research participation often address whether payment motivates participation at the expense of a full risk appraisal (Dickert and Grady 1999, Grady 2001, Halpern et al. 2004, Sears 2001, Tishler and Bartholomae 2002). O'Neill (2003) argues that the ethical purpose of informed consent is to ensure that research participants are not deceived or coerced and therefore payment for participation would suggest that individuals may have been influenced in their decision-making. US federal guidelines on payments to research subjects state that payments should not constitute inducement or coercion, and should not be portrayed by researchers as a personal benefit of participation (US Food and Drug Administration 1998, US Office of Human Subjects Research 2000). Thus, researchers face substantial challenges in attempting to set fair and reasonable amounts that would not be viewed as undue inducement to participate (Deschênes et al. 2001: 230). Meanwhile, studies have shown that healthy individuals are often motivated by payment (Bigorra and Baños 1990, Cunny and Miller 1994, van Gelderen et al. 1993), and think of money as a benefit of participation (Resnick 2008).

Our study also offers insight into decision-making as it relates to payment of participants. Healthy volunteers, whom Grady (2005: 1685) calls 'independent contractors in research', are often motivated by money to participate in studies, especially since participation usually involves little or no personal benefit. While Grady (2005: 1683) asks whether 'financial incentives blind potential participants to the risks of research when making decisions regarding participation', our data suggest that money is only part of the answer. When coupled with sufficient levels of trust, especially when bolstered by positive past experiences, money appears to hold center stage because any concerns have been relegated to a proxy – the IRB, Certificate of Confidentiality, or the research organization.

One limitation of our study is that interviews did not occur immediately after consent (or at the point when decliners decided not to join). While it could be argued that this accounts for poor understanding of the key features, a survey of 924 participants in another US biobank – where the median time since enrollment was approximately one year – found that poor understanding of study features was due to a lack of comprehension rather than memory loss because time since participation was not related to the percentage of correct responses (McCarty et al. 2007). Other studies conducted immediately after consent have also shown poor understanding among participants (Fortun et al. 2008, Barr 2006, Dixon-Woods et al. 2007).

Ours was a sophisticated sample – almost all were college educated with work experience in healthcare, and had participated in other research studies; some talked of regularly seeking research studies. In fact, membership in a research university likely makes people aware of studies that pay participants, which may have influenced our finding regarding the importance of payment. Another study on informed consent conducted with educated healthy volunteers (over half of whom were clinical medical students) found that immediately after consent very few could name key study features but virtually all knew the exact amount of payment paid for participating (Fortun et al. 2008).

This study adds to our understanding of participation in biobanks in important ways. While biobanking has many ethical dimensions, one focus is on how individuals who volunteer their samples understand the fundamental contribution they make to a biobank's success – an understanding that typically begins with a researcher's invitation to participate. We have highlighted the surprising similarities of joiners and decliners. Like decliners, joiners made decisions outside of any formal enrollment process, by independent means that may or may not have included available study resources. Also, payment was important for both groups. For joiners, it was associated with a short-term focus framing risk and benefit: one could get money now, trusting existing safeguards to offer necessary protection. In contrast, decliners, despite their interest in payment, were not convinced that those same safeguards would prove effective over time.

In conclusion, two points seem particularly important to consider for their implications in designing biobank enrollment processes. One relates to trust in external safeguards – the IRB, Certificates of Confidentiality, and the research institution – that played such an important role in assuring joiners of their personal privacy and confidentiality. While these safeguards should receive more emphasis and description in recruitment and enrollment materials, the scope of their protection should not be overstated given the uncertainties related to the use and storage of samples for future genetic research. For example, it may be difficult to discern the full extent of relevant federal, academic, and private interests associated with a biobank for its entire lifespan. Participants' reliance on the Certificate of Confidentiality may be particularly worrisome due to the current lack of evidence of their full legal effect (Beskow, Dame and Costello 2008). In fact, some have cautioned that because the guarantees associated with Certificates are not definitive, investigators are 'obligated' to inform potential participants of their limitations (Catania et al. 2007). Consent processes are challenged to make clear these complexities. The second point concerns the influence of subject payment on participation in biobanks. Given its motivating appeal, the careful presentation of study risks and benefits, as well as the safeguards described above, are all the more important. The growth of biobanks will rely upon the contributions of healthy volunteers. Accurate portrayal of the information healthy volunteers find most relevant will promote their thoughtful appraisal of participation.

Acknowledgments

Thank you to the investigators of the Environmental Polymorphisms Registry for their time and information. We are very grateful to Gail Henderson, Nancy King, Robert Mitchell, and Rene Sterling for providing comments and suggestions on drafts of this chapter. Special thanks also to Michele Easter for her work on the interview project in its earliest stages. The project was supported by Grant Number P50HG004488 from the United States National Human Genome Research Institute. The content is solely our responsibility and does not necessarily represent the official views of the National Human Genome Research Institute or the National Institutes of Health.

Bibliography

Anderlik, M.R. and Rothstein, M.A. 2001. Privacy and confidentiality of genetic information: What rules for the new science? *Annual Review of Genomics and Human Genetics*, 2, 401-33.

Barr, M. 2006. 'I'm not really read up on genetics': Biobanks and the social context of informed consent. *BioSocieties*, 1(2), 251-62.

Bates, B.R., Lynch, J.A., Bevan, J.L. and Condit, C.M. 2005. Warranted concerns, warranted outlooks: A focus group study of public understandings of genetic research. *Social Science & Medicine*, 60(2), 331-44.

Beskow, L.M., Dame, L. and Costello, E.J. 2008. Research ethics: Certificates of Confidentiality and compelled disclosure of data. *Science*, 322(5904), 1054-55.

Bigorra, J. and Baños, J.E. 1990. Weight of financial reward in the decision by medical students and experienced healthy volunteers to participate in clinical trials. *European Journal of Clinical Pharmacology*, 38(5), 443-46.

Boulton, M. and Parker, M. 2007. Informed consent in a changing environment. *Social Science and Medicine*, 65(11), 2187-198.

Busby, H. 2004. Blood donation for genetic research: What can we learn from donors' narratives? in *Genetic Databases: Socio-ethical Issues in the Collection and Use of DNA*, edited by R. Tutton and O. Corrigan. London: Routledge, 39-56.

Catania, J.A., Wolf, L.E., Wertleib, S., Lo, B. and Henne, J. 2007. Research participants' perceptions of the Certificate of Confidentiality's assurances and limitations. *Journal of Empirical Research on Human Research Ethics*, 2(4), 53-9.

Caulfield, T. 2007. Biobanks and blanket consent: The proper place of the public good and public perception rationales. *King's Law Journal*, 18(2), 209-26.

Chulada, P.C., Vahdat, H.L., Sharp, R.R., DeLozier, T.C., Watkins, P.B., Pusek, S.N. and Blackshear, P.J. 2008. The Environmental Polymorphisms Registry: A DNA resource to study genetic susceptibility loci. *Human Genetics*, 123(2), 207-14.

Churchill, L.R., Nelson, D.K., Henderson, G.E., King, N.M.P., Davis, A.M., Leahey, E. and Wilfond, B.S. 2003. Assessing benefits in clinical research: Why diversity in benefit assessment can be risky. *IRB: Ethics and Human Research*, 25(3), 1-8.

Coleman, C.H., Menikoff, J.A., Goldner, J.A. and Dubler, N.N. 2005. *The Ethics and Regulation of Research with Human Subjects*. Newark, NJ: LexisNexis.

Corrigan, O. 2003. Empty ethics: The problem with informed consent. *Sociology of Health & Illness*, 25(7), 768-92.

Cunny, K.A. and Miller, H.W. 1994. Participation in clinical drug studies: Motivations and barriers. *Clinical Therapeutics*, 16(2), 273-82.

Deschênes, M., Cardinal, G., Knoppers, B.M. and Glass, K.C. 2001. Human genetic research, DNA banking and consent: A question of 'form'? *Clinical Genetics*, 59(4), 221-39.

Dickert, N. and Grady, C. 1999. What's the price of a research subject? Approaches to payment for research participation. *New England Journal of Medicine*, 341(3), 198-203.

Dixon-Woods, M., Ashcroft, R.E., Jackson, C.J., Tobin, M.D., Kivits, J., Burton, P.R. and Samani, N.J. 2007. Beyond 'misunderstanding': Written information and decisions about taking part in a genetic epidemiology study. *Social Science & Medicine*, 65(11), 2212-222.

Fortun, P., West, J., Chalkley, L., Shonde, A. and Hawkey, C. 2008. Recall of informed consent information by healthy volunteers in clinical trials. *QJM*, 101(8), 625-29.

Gottweis, H. 2002. Gene therapy and the public: A matter of trust. *Gene Therapy*, 9(11), 667-69.

Grady, C. 2001. Money for research participation: Does it jeopardize informed consent? *American Journal of Bioethics*, 1(2), 40-4.

Grady, C. 2005. Payment of clinical research subjects. *The Journal of Clinical Investigation*, 115(7), 1681-87.

Haimes, E. and Whong-Barr, M. 2004. Levels and styles of participation in genetic databases: A case study of the North Cumbria Community Genetics Project, in *Genetic Databases: Socio-ethical Issues in the Collection and Use of DNA*, edited by R. Tutton and O. Corrigan. London: Routledge, 57-77.

Halpern, S.D., Karlawish, J.H.T., Casarett, D., Berlin, J.A. and Asch, D.A. 2004. Empirical assessment of whether moderate payments are undue or unjust inducements for participation in clinical trials. *Archives of Internal Medicine*, 164(7), 801-803.

Hansson, M.G., Dillner, J., Bartram, C.R., Carlson, J.A. and Helgesson, G. 2006. Should donors be allowed to give broad consent to future biobank research? *Lancet Oncology*, 7(3), 266-69.

Hoeyer, K. 2003. 'Science is really needed – that's all I know': Informed consent and the non-verbal practices of collecting blood for genetic research in northern Sweden. *New Genetics and Society*, 22(3), 229-44.

Hoeyer, K. 2004. Ambiguous gifts: Public anxiety, informed consent and biobanks, in *Genetic Databases: Socio-ethical Issues in the Collection and Use of DNA*, edited by R. Tutton and O. Corrigan. London: Routledge, 97-116.

Hoeyer, K. and Lynöe, N. 2006. Motivating donors to genetic research? Anthropological reasons to rethink the role of informed consent. *Medicine, Health Care and Philosophy*, 9(3), 13-23.

McCarty, C.A., Nair, A., Austin, D.M. and Giampietro, P.F. 2007. Informed consent and subject motivation to participate in a large, population-based genomics study: The Marshfield Clinic Personalized Medicine Research Project. *Community Genetics*, 10(1), 2-9.

Miller, T. and Boulton, M. 2007. Changing constructions of informed consent: Qualitative research and complex social worlds. *Social Science & Medicine*, 65(11), 2199-211.

O'Neill, O. 2003. Some limits of informed consent. *Journal of Medical Ethics*, 29(1), 4-7.

Resnick, D.B. 2008. Increasing the amount of payment to research subjects. *Journal of Medical Ethics*, 34(9), e14.

Sears, J.M. 2001. The payment of research subjects: Ethical concerns. *Oncology Nurses Forum*, 28(4), 657-63.

Tambor, E.S., Bernhardt, B.A., Rodgers, J., Holtzman, N.A. and Geller, G. 2002. Mapping the human genome: An assessment of media coverage and public reaction. *Genetics in Medicine*, 4(1), 31-6.

Tishler, C.L. and Bartholomae, S. 2002. The recruitment of normal healthy volunteers: A review of the literature on the use of financial incentives. *Journal of Clinical Pharmacology*, 42(4), 365-75.

United States Department of Health and Human Services. 2003. *Certificates of Confidentiality: Background Information*. [Online] Available at: http://grants.nih.gov/grants/policy/coc/background.htm [accessed 14 October 2008].

United States Department of Health and Human Services. 2005. *Protection of Human Subjects. US Code of Federal Regulations, title 45, part 46*. [Online] Available at: http://www.hhs.gov/ohrp/humansubjects/guidance/45cfr46.htm#46.116 [accessed 13 October 2008].

United States Department of Health and Human Services. 2008. *Office of Civil Rights – HIPAA*. [Online] Available at: http://www.hhs.gov/ocr/hipaa/ [accessed 30 September 2008].

United States Food and Drug Administration. 1998. *Payment to Research Subjects*. [Online] Available at: http://www.fda.gov/oc/ohrt/irbs/toc4.html#payment [accessed 16 October 2008].

United States National Cancer Institute. 2007. *National Cancer Institute Best Practices for Biospecimen Resources*. Bethesda, MD: NCI, NIH. [Online] Available at: http://biospecimens.cancer.gov/global/pdfs/NCI_Best_Practices_060507.pdf [accessed 27 February 2009].

United States National Institute of Environmental Health Sciences – National Institutes of Health. 2006. *The Environmental Polymorphisms Registry*. [Online] Available at: http://www.niehs.nih.gov/research/resources/databases/epr/ [accessed 30 September 2008].

United States Office of Human Subjects Research. 2000. *Information Sheet 20*. [Online] Available at: http://ohsr.od.nih.gov/info/sheet20.html [accessed 16 October 2008].

van Gelderen, C.E.M., Savelkoul, T.J.F., van Dokkum, W. and Meulenbelt, J. 1993. Motives and perception of healthy volunteers who participate in experiments. *European Journal of Clinical Pharmacology*, 45(1), 15-19.

Wendler, D. 2006. One-time consent for research on biological samples. *British Medical Journal*, 332(7240), 544-47.

Whong-Barr, M. and Haimes, E. 2003. Why say no? Reasons for non-participation in the North Cumbria Community Genetics Project. *European Journal of Human Genetics*, 11(Supp. 1), 33-9.

PART 3
Privacy and Access

Chapter 9
Privacy Interests in Biobanking: A Preliminary View on a European Perspective

David Townend, Mark J. Taylor, Jessica Wright and Dita Wickins-Drazilova

There is certainly an interest in biobanking and research using genetic information in Europe. It offers great possibilities in health care, both at the level of understanding disease and potentially enabling the production of new and more effective therapies and diagnoses, and so it has implications for European commerce and public health. Europe has a developing tradition in this field, of protecting the fundamental rights and freedoms of its citizens, for example through data protection.[1] However, the new genetic data processing possibilities are in potential conflict with these fundamental rights, because the real research value of biobanking and research using genetic data will be in its relation to the medical and environmental life-story of the data subject. Further, the information contained in the samples is not simply information about the donor data subject; at the same time it discloses information about others who are genetically related to the donor subject. The nature of this information makes it not only attractive for a new generation of medical researchers, but also for a new generation of policy/decision-makers. It allows the potential for understanding the origins of disease, and promises more tailored drug regimes, but at the same time in other hands it allows for risk assessments which could be used to express preferences between individuals or groups of individuals in employment or insurance. It could be the basis of positive medical advance, and of discriminatory social retreat. It is therefore a matter for regulation.

Legislators, when considering any new law or regulation but particularly in an area as sensitive and highly charged as the regulation of genetic information and biobanking, have to decide how far they should seek to lead public opinion in the creation of the regulatory regime and how far they should simply seek to reflect public concerns when regulating the area. While few would argue that it is the job of regulators unthinkingly to translate populist sentiment into law, it

1 See EC Directive 95/46/EC, and also the Charter of Fundamental Rights of the European Union (2000) which has not only a right to private and family life (Article 7), but a stand-alone right to the protection of an individual's personal data (Article 8). The Charter is not yet in force following the failure of the Member States to accept the Draft Treaty Establishing a Constitution for Europe (2004) and the Treaty of Lisbon (2007).

is nevertheless important that regulation, and regulators, maintain public trust. This is in part influenced by the nature of the broad genetic project: given that biobanking and research with genetic information works best with the inclusion of large numbers of people from diverse backgrounds, how can the regulation of this area best reflect the expressed opinions of citizens in order to generate sufficient trust and confidence, and thereby encourage their participation? At what level should such regulation operate? National, European, International? There may be operational advantages to harmonized regulation, but, to what extent might this overlook national perspectives? The adequacy of European regulation will be determined, in part, by its ability to recognize and accommodate 'local' interests in research using genetic databases and biobanks.[2]

Privacy Interests and the Law Relating to Biobanks and Genetic Databases

This chapter will first provide a brief overview of the PRIVILEGED project, an outline of the results of the first stage of the project with a focus on potential privacy interests concerning the control of data used in research, and preliminary conclusions on reflecting the diverse interests expressed relating to consent within regulation and the lack of a distinctive 'European' perspective on privacy. Our preliminary conclusions may be traced through the following chain of reasoning. There are only limited surveys of public opinion on the issue of privacy and data protection in relation to genetic information and biobanking. Even within this limited range, a spectrum of interests in research using biobanks and genetic databases has already been expressed by the general public and others. The characterization of such interests as 'privacy interests' depends upon an underlying concept of privacy. It is possible to construct a 'European' concept of privacy relying upon existing legal norms expressed through the Council of Europe, its legal instruments and institutions.[3] Such a concept would characterize certain of the interests expressed as 'privacy interests' and would, by definition, inevitably offer those interests some degree of legal protection. The adequacy of the concept of privacy currently underlying existing legal norms ought not, however, to be assumed. The concept of privacy is a contested concept and there is a range of alternative views expressed within the literature. Alternative concepts of privacy would accommodate the interests that have been expressed in research using genetic databases and biobanks as privacy interests to varying degrees. Some variation in the accommodation of particular interests as 'privacy interests'

2 We can imagine here, amongst 'local' interests, perhaps the interests of minority groups, such as the Sami people in Norway, but also national perspectives on the relative significance of particular interests that are recognized to be shared with others but valued more or less highly than in other countries.

3 See the European Convention on Human Rights (1950), specifically Article 8: The right to respect for private and family life.

may represent differences of emphasis between alternative concepts of privacy, but, some variations may represent more fundamental differences. *Some* of the interests that have been expressed in relation to research using genetic databases and biobanks may well resonate more closely with a concept of privacy *dissimilar* to the one that current legal norms imply. This may be important. If there is a desire to maintain public trust in the mechanisms of regulation and governance, then the adequacy of any legal concept of privacy to capture interests expressed in research is of concern. The relevant question facing those responsible for shaping the law is not simply: How ought we to *balance* a claim to privacy against conflicting interests? There is a prior, fundamental question about which concept of privacy ought we to be using to understand legal claims to privacy in the first place? Altering the underlying concept may shift the perceived parameters of any *legal* claim to privacy. This may do more than simply shift the balance between a right to privacy and competing claims (such as a public interest in research). It may capture (or exclude) certain kinds of interests expressed (regarding research using genetic databases and biobanks) as 'privacy interests' not at all susceptible to protection under the law.

Privacy itself goes to supporting particular kinds of relations between persons, and between persons and the state, and that itself helps to constitute a particular kind of society.[4] This can lead to the suggestion that a single interpretation of privacy should be favoured. For example, in the UK House of Lords judgment in *R (S and Marper) v. Chief Constable of the South Yorkshire Police* ([2004] UKHL 39: para 27), Lord Steyn held that the question of whether the right to a private and family life[5] was engaged (in particular circumstances) 'should receive a uniform interpretation throughout member states, unaffected by different cultural traditions'.[6] If such an approach is accepted, those responsible for the regulation and governance of biobanks and genetic databases should not underestimate the significance of getting 'right' the underlying concept of privacy to be employed when establishing the 'privacy claims' of those whose information may be used for genetic research. If it is true that Europe must adopt *a uniform* concept of privacy, then, we should at least be fully appreciative of the relative strengths of the alternatives. Indeed, this is the case if a range of concepts is accommodated.

Equally, if we wish to maintain public trust in biobanks, it seems clear that we must first be aware of the interests that individuals perceive to be vulnerable.

---

4 We find ourselves in some sympathy with the view expressed by the fictional character Howard Roark: 'Civilization is the progress toward a society of privacy … Civilization is the process of setting man free from men' (Rand 1943). However, for a rather more balanced view of privacy as a social issue see Margulis (2003).

5 Provided for by Article 8(1) of the European Convention on Human Rights (1950).

6 Although his Lordship *did* also find that, 'when one moves on to consider the question of objective justification under article 8(2) the cultural traditions in the United Kingdom are material' (para 27). While this undoubtedly complicates matters, it doesn't affect the point that we are seeking to advance.

A decision must then be taken on how best to protect these interests (in the face of potentially competing claims). The law on privacy is only one mechanism capable of providing such protection, but, it is an important one. Understanding the range of interests that *could* be protected through the law of privacy, and making an informed choice between the alternatives, requires an understanding of two different spectra of possibilities: the range of possible (perceived) interests themselves and the range of alternative possible concepts of privacy. Together they tell of the particular concepts of privacy that might (or might not) capture the specific interests that individuals have expressed.[7]

Within this chapter we do not seek to provide an answer to the question: Which concept of privacy ought to be adopted? Indeed, we do not even seek here to fully defend the claim that existing legal norms reflect a particular concept of privacy incapable of accommodating the broadest range of possible interests. Some of these issues have been addressed by us elsewhere (in particular see Taylor 2006, and 2007). Comprehensive consideration of them all will have to wait until the PRIVILEGED project is more advanced and we have rather more space to pursue them. Here, we seek only to illustrate the point that different interests might be recognized as 'privacy interests' by alternative concepts of privacy. In doing so, however, we do draw preliminary conclusions about the available survey information, and whether the interests reported in the surveys and other available literature indicates a distinctive 'European' perspective on privacy in relation to research using genetic data and biobanking.

The PRIVILEGED Project

PRIVILEGED is a three-year European Commission Framework Programme 6 Coordination Action project. A 'Coordination Action' is a funding vehicle to bring together individuals (perhaps across different disciplines) to discuss their existing work and conclusions about a particular area and, through that discussion and work, to develop and disseminate new and greater understandings and connections through the data. It is not primarily a funding vehicle for new research. PRIVILEGED, therefore, brings together academic lawyers, medical practitioners and researchers, ethicists and sociologists from all the Member States of the EU, Norway and Iceland (from the European Economic Area), and, by way of contrast with the European group, Israel, Japan and Taiwan. The project's aim is to present to the Commission a series of regulatory choices concerning privacy protection (with particular consideration of data protection in such protection) in relation to processing of genetic information, particularly in research biobanking. In order to achieve this, the project has three phases. First, the members undertook a literature search of reports of surveys of citizens' expressions about their interests in

7 Our present concern is those interests expressed with respect to research using biobanks and genetic data, but the issues raised are of broader concern.

privacy,[8] particularly in relation to their genetic information and biobanking. This was not new research, but a survey of the reported research of others in each of the states represented in PRIVILEGED. In the second stage of the work (the current stage at the time of writing this chapter), the existing regulatory frameworks at the European and national levels will be analysed, assessing the type of privacy interests that they seek to protect. The final part of the work builds on the findings and materials from the first two parts of the project in an exploration of how the citizens' expressed interests can be better accommodated within the law relating to research using genetic information and biobanking. This will be done by presenting a range of regulatory possibilities that achieve different privacy landscapes, in particular examining how the data protection regime (especially the Directive 95/46/EC) can be arranged to optimize both research and privacy protection. Within this part of the project we will examine areas which present particular difficulties in the existing regulatory regimes (for example, the assumption that anonymization of data or informed consent necessarily offers optimum protection in this arena). One of the major themes in the research is the tension between the individual's rights and the shared quality of genetic information (i.e. individuals share their genetic make-up with others – most significantly with family members, ethnic groups and those with the same or similar genetic disorders or conditions). A focus of the project is to find ways of accommodating these competing rights.

The Views of European Citizens Towards Privacy, Genetic Information and Biobanking

The first part of PRIVILEGED is a literature review by the members of the project. The literature identified falls into two types: European and national. From this, we concluded that there have been limited systematic, Europe-wide studies asking citizens to identify their privacy (and other) interests in their personal information, which are relevant to genetic research and research biobanking. There have been two Eurobarometer[9] surveys containing relevant questions, but within the context

8 Here we were simply looking to understand what people seemed to attribute value to, or raise an expectation with respect to, when considering the acquisition or use of data. If they expressed a view that X or Y ought to be done, then we considered that they perceived an interest in X or Y *being* done. In this way we sought to cast the net wide and to capture what may have been either 'self interests' or 'moral interests'. The interest itself may not have been directly expressed (or even considered) by the individual expressing an opinion. The opinion or attitude, however, we have taken to constitute an 'expression of interest' and to represent, collectively, a range of 'expressed interests'.

9 Eurobarometer are quantitative surveys used by the European Commission to obtain an indication of public opinion. The Special Eurobarometers mentioned in this chapter were based on interviews with approximately 1,000 people in each country (or with a figure proportional to the country size). Each Eurobarometer may cover different countries in the wider EEA. However, figures quoted for Europeans do not include the EEA countries.

of broader research questions. *Social Values, Science and Technology* (European Commission 2005) asked participants in question 15a.3 how important protecting information about one's private life from misuse and exploitation would be for society in 10 years time? In question 17, it asked about attitudes towards applications of sciences and technology that may develop over the next 20 years (including storing everyone's genetic data for the purposes of catching criminals or to study the genetic causes of human diseases; as well as genetic testing either to identify one's genetic disease profile (both those diseases with and without known cures) or to identify children's talents and weaknesses). Across the European countries surveyed, 94 per cent of people indicate that they believe that protecting personal data will be important in the next 10 years (with 67 per cent indicating that it would be very important, a majority position in all countries).[10] There is evidence of a suspicion that research using genetic databases or biobanks might not be consistent with the strong protection of personal data desired. Twenty-one per cent of Europeans would never approve of 'storing everyone's genetic data so that criminals could be caught easily', 34 per cent would never approve of '[d]eveloping for everybody a genetic test that would tell us about diseases we might get, even if we cannot do anything about them', and only 6 per cent approve (in all circumstances, without regulation) of the development of a genetic test for the identification of talents and weaknesses in children. However, there is greater approval for 'Storing all the genetic data of our population in data banks in order to study the genetic causes of human diseases', with 17 per cent of people surveyed responding that they would 'never agree'. The more positive attitude towards some sorts of medical research could be very important for biobanks and genetic databases engaged in this research.

In *Europeans and Biotechnology* (Gaskell et al. 2006: chapter 5.3), five questions were asked about the respondents' opinions towards the use of genetic data. Again, the results show a variety of responses. For example, 58 per cent of Europeans indicated that they would be prepared to allow their genetic data to be placed into a 'national data bank for research into the origins of diseases', and 59 per cent of respondents across Europe were willing to allow the police *access* to people's genetic information to solve crime (compared to 33 per cent supporting *storage* of everybody's genetic data for forensic purposes in the previous survey). Such differences need to be explored further, using qualitative methods to tease out why there is a difference between access and storage in this example. The *Europeans and Biotechnology* survey indicates that people were less willing to allow government social security agencies to access people's genetic information than the police, and again less likely to allow private insurance companies access. This survey indicated that 64 per cent of Europeans would take a test to identify any serious genetic diseases that they might get.

10 This could be weighed against the 95 per cent that believe that protecting the freedom of speech and information will be important in 10 years time.

These surveys are valuable as they provide Europe-wide information about citizens' attitudes towards the use of genetic information and, to some extent, about people's fears about the misuses of such data. The Eurobarometer studies also disclose national differences in the response rates, enabling some degree of comparison to take place. It is, for example, interesting to contrast the responses to the use of genetic information in Iceland and the UK, given that both have a prominent but significantly different biobank. The UK and Iceland, in some questions, sit at the two outside edges of the responses, with a greater acceptance in the UK in some of the questions and a greater concern in Iceland in others.[11]

Overall, however, these Eurobarometer questions are only a small number of quantitative questions in studies that were not directly focused on gathering information exploring the possibilities of creating legislation to protect interests in genetic information and biobanking. A forthcoming Eurobarometer study in late 2009 will address key questions about attitudes to genetic information and biobanking. This must be welcomed, and will contribute towards the body of strong Europe-wide information now beginning to emerge.

However, the PRIVILEGED literature review (even with the presence on the horizon of a new Eurobarometer study in the area), shows that there is a significant absence of detailed and systematic, Europe-wide, *qualitative* studies of citizens' attitudes towards privacy in genetic information and biobanking. In particular, the project's review of the surveys indicates that there is still a great need for more studies which identify individual citizen's attitudes to the group-quality of genetic data. If such surveys are important for the development of effective, trust-building legislation in the area (which we believe they are) and the current and foreseeable Eurobarometer work is not yet sufficient to provide evidence which could support legislative choices (which we believe it is not) then the next question is: is there a body of national surveys which fit together to form a significant and coherent body of detailed information to meet the gap in the Europe-wide survey landscape?

Before further reporting on the literature review, we must sound a note of caution about the information. Here we are trying to show something of the range of studies that have been undertaken, and to give a basic indication of the sorts of responses that people give. This is not a full, systematic report of the literature review. It does not suggest that the methodologies of the different surveys are comparable, indeed, each survey has its own independent methodology and these make the surveys very difficult to compare. Likewise, we are not seeking here to

11 Two examples from Q17 of *Social Values, Science and Technology* (European Commission 2005) illustrate this. The first is on approval of storing everyone's genetic data so that criminals can be caught easily (in Iceland 43 per cent would never approve, in the UK only 13 per cent feel the same), and the second on approval for development of a genetic test that would tell us about diseases we might get (in Iceland 45 per cent would never approve, in the UK, only 23 per cent feel the same).

make any comment about the validity of the studies. Here the purpose is to show something of the sorts of questions asked and the range of responses that they provoke.

From the PRIVILEGED literature review, there is a great variation between the European Member States[12] in relation to the number and scope of surveys[13] of citizen's interests in privacy in research using genetic information and biobanking. In some countries where large-scale biobanking projects had been undertaken for research purposes, a range of public opinion surveys have been conducted on the topic.[14] In a second group, there were a small number of countries (Norway, Sweden, Austria, France and the UK) where individual academics with interests in the area had produced a large amount of work.[15] In another group there may not be large-scale projects, but public interests have still been surveyed to some extent, or at least on related topics.[16] However, in other states PRIVILEGED participants

---

12 And with the other states represented in the project (Norway, Iceland, Israel, Japan and Taiwan).

13 The term survey is used to describe research conducted to ascertain public opinion using quantitative and/or qualitative methods. The types of surveys include telephone or face-to-face interviews, focus groups, postal (or online) questionnaires, citizen's conferences and open public consultations. In the project we also considered organizational reports, work by anthropologists, reviews of media discourse (and public debate), discussion around legislative changes and Court cases.

14 In Estonia and Iceland the only reported national survey of public opinion on the genome projects were conducted in 2002 within the international research project ELSAGEN (Ethical, Legal and Social Aspects of Human Genetic Databases. A European Comparison). In Estonia they interviewed a representation sample of 917 respondents face-to-face (see Korts 2004). In Iceland, a telephone interview was conducted with 1500 people (see Gudmundsdóttir et al. 2007). In England there have been interviews (both face-to-face and telephone), focus groups, postal surveys as well as one public consultation. These have been conducted both by the funders of UKBiobank and independently, include both qualitative and quantitative methods and focus on both the general public as well as specific donor populations or disease groups. Relevant surveys have also been conducted in Scotland (for the Generation Scotland project, see Haddow et al. 2004, who conducted focus groups and interviews). In Norway studies have been conducted in conjunction with the Nord-Trøndelag Health Study (HUNT) (see the focus group study in Skolbekken et al. 2005). In Sweden a great number of studies have been conducted (both by postal surveys and interviews) (see note 22 below). See the PRIVILEGED project webpages (www.privilegedproject.eu) for more details on the studies conducted in these countries.

15 For example, the work of Klaus Hoeyer, Helen Busby, Richard Tutton, Ulrike Felt, Pascal Doucernou, Marcel Bister, Lars Ursin, John-Arne Skolbekken, Ole Brekke, Lene Ring, Åsa Kettis, and Ingrid Metzler.

16 Austria (the Austrian Council for Research and Technology Development (2003) organized a citizen's conference on 'Genetic data: where from, where to, what for?'), Belgium (a citizens' conference run by the King Badouin Foundation (2003) on 'Is It In My Genes?'), Denmark (interviews with research participants in Lind et al. 2007,

reported that very little work has been undertaken in this area, or, surprisingly on attitudes to privacy generally.[17] There are, therefore, quite a number of surveys relevant to the area with a number of interesting methodologies and results, and in countries where such material is available, it can clearly influence the creation of domestic legislation. However, the difference in methodologies contributes towards the considerable difficulty in comparing sociological research across countries. Therefore, we cannot say, following our work in the first stage of PRIVILEGED, that there is a systematic, detailed body of survey work that can be read together as a sufficient map of the detailed opinions of European citizens about their privacy interests in relation to research using genetic information and biobanking. Having said that the work is insufficient and rather variable between the different countries represented in PRIVILEGED, there are many valuable and informative surveys from which opinions relevant to potential regulation in this area can be drawn: the cupboard is far from bare, and the results should inspire further work to explore how far the findings are similar across Europe, and then how far there is a distinctive European attitude towards privacy in relation to genetic information and biobanking. PRIVILEGED has surveys which present a range of potential concerns from citizens. However, it has not found surveys that show that these concerns are present in different parts of the EU, or are the full range of concerns in the EU. Therefore, we conclude that the surveys do not yet provide a sufficient evidence platform upon which to base European-wide legislation, confident that it takes into account the full range of expressed interests, or to make choices for particular privacy regimes in the area that could be said to have considered the full range of interests expressed by the public. There is a significant research gap open at the present time if legislating on the basis of a full understanding of the public's expressed concerns, aspirations and goals is important.

'Participation in environmental health research by placenta donation – a perception study'), Finland (Tupasela et al. 2007 on the perceptions of Finns towards the biomedical use of tissue collections and biobanks), Germany (the Stiftung Deutsches Hygiene-Museum held a Citizen's Conference on Genetic Testing in 2001), The Netherlands (interviews and focus groups in the Public Survey Genomics in 2002 and 2005 (see Stichting Consument en Biotechnologie 2002, Pin and Gutteling 2005), The Republic of Ireland (see Cousins et al. 2005, who conducted telephone surveys of 2,294 people on biomedical research) and Spain.

17 This is, of course, the most difficult to reference. However, Members in Bulgaria, Cyprus. Czech Republic, Greece, Hungary, Italy, Latvia, Lithuania (although there is a survey forthcoming), Malta, Poland and Portugal have been unable to find significant work in their domestic literature. This is despite the existence of Biobank Hungary and the Genome Database of the Latvian Population.

Citizens' Interests in Privacy in Research Using Genetic Information and Biobanks Reported in National Surveys

Having said that the platform is rather variable between different countries represented in PRIVILEGED and insufficient for Europe-wide legislation, there are many valuable and informative surveys from which some conclusions about the nature of regulation for this area can be drawn. This information indicates results that should inspire further work to explore how far the findings are similar across Europe, and then how far there is a distinctive European attitude towards privacy in relation to genetic information and biobanking. The surveys also draw us to a conclusion for regulators in the area, concerning the treatment of a variety of different and potentially contradictory public opinions when developing regulations that maximize public trust, confidence and participation. In this section, and by way of an example of the range of data already available, we will draw on one line in the available surveys that can be said to show a desire for control over the data. One of the major concerns that we identify in the current landscape of privacy and data protection framework for the regulation of genetic information and biobanks is the relationship between the individual right and the group right in the data. At the heart of this tension is the ability to control information. In the surveys, there are many interesting questions relating to opinions on the use of the data, (informed) consent and the right to know or not to know results flowing from research. Whilst there are other lines of interest which are emerging from studying the data, and these will be explored elsewhere,[18] we have chosen this line for comment here because it will form the basis for comment about the relationship between citizens' opinions and legislation later in this chapter.

Whilst the Eurobarometer surveys show an acceptance of the social benefit of medical research which studies the cause of diseases, at the same time, some citizens express concern in national surveys about genetic research in general, or about the uses to which the data or samples could be put.[19] Equally and quite obviously, others within the same surveys do not share the same difficulties. There are also some surveys that report that some participants see the potential for discrimination

18 For a more thorough treatment of the expressed interests, see further publications from the PRIVILEGED project online at: www.privilegedproject.eu.

19 For example, in the UK MORI Social Research (2001) 33 per cent of a people's panel surveyed agreed with the statement that human genetics research is 'tampering with nature and is therefore unethical'. In Calnan et al. (2005) 52 per cent of a postal survey of the UK general public agreed with the statement, and in a telephone survey of the general public in the Republic of Ireland by Cousins et al. (2005) 42 per cent agreed that 'research on human genetics is tampering with nature'. An example on opinions relating to the use of research results is that 32.8 per cent of the respondents in Hoeyer et al. (2005) (donors to a Swedish biobank) saw it as important that the results of the research were not used to offer selective abortion.

and misuse of the data.[20] This is especially the case in relation to commercial uses of the data, particularly in relation to insurance.[21] Again, however, each survey also shows that there are participants who do not share these views and give the opposite response to the questions. Participants in some surveys express concerns about the commercialization of the data and products emerging from the use of the data.[22] Again, others within the studies express the opposite opinion. This broad range of expressed views about the use to which the data is put show a desire for some level of control of disclosed data (so called 'reach-through provisions') for some people.

There are other issues of control that are expressed in relation to the gathering of the data and the relationship between the research subject and his or her genetic information. Informed consent is an important safeguard to many participants in a range of studies.[23] However, a number of different positions in relation to consent are reported by different surveys. These range from a requirement of informed consent required for each piece of research on the data, whether it be for the original research for which the data were gathered or for subsequent new research,[24] to satisfaction with more general or blanket consent,[25] to reports of the opinion that no consent is seen as necessary where data is gathered for medical

20 See, for example, the report of the Belgian citizen's conference *Is it in My Genes?* which expressed the concern that over the long-term genetic testing may lead to stigmatization of ill persons or persons with a handicap (King Baudouin Foundation 2003). In UK focus groups, a fear of losing control of genetic information due to the number of interested parties was expressed (Levitt and Weldon 2005).

21 See, for example, MORI Social Research (2001) where 80 per cent of respondents (UK general public) felt that genetic information should not be used to set insurance premiums. In the Icelandic general population, only 16 per cent of the respondents felt that insurers should have access to medical databases containing information on medical records and biological samples (Gudmundsdóttir and Nordal 2007).

22 For example, in Skolbekken et al. (2005) focus groups related to a genetic research project in Norway were concerned that the biobank should not be commercialized, especially by the pharmaceutical industry. In Hoeyer et al. (2005) 54.9 per cent of the respondents (from donors to a Swedish biobank) indicated that it was important that commercial interests did not determine the research outlook. However, Richards et al. (2003), for example, interviewed donors in a UK genetic research project and found that they had few concerns about commercial companies having access to samples.

23 For example, in Eriksson's study (2007) 94 per cent of the respondents to the postal survey of the general public believed that informed consent should be obtained before genetic testing was undertaken.

24 For example, Gudmundsdóttir and Nordal (2007) found that 57 per cent of their respondents in the Icelandic general public agreed that individuals should be re-contacted for each new collection of biological material for use in genetic research.

25 For example, in the survey in Finland by Tupasela et al. (2007) 22 per cent of the sample indicated that they would be satisfied with giving open consent. The same study found that 44 per cent wished to create their own consent formula when samples were to be transferred to a biobank.

research.[26] The surveys also ask questions which show that there can be different opinions in the case of further use of data originally collected for another purpose for medical research, especially in the case where samples could be seen as 'surgical waste'.[27]

One element of control that may be granted to people is the ability to choose between different options relating to the use of the data (often in the form of 'consent options'). In research using genetic data, important consent options may concern, for example, the 'right to know' and the 'right not to know'. Research using genetic information and biobanking has the potential (if the data remains identifiable to its donor) to reveal new information about the donors. The nature of that information is that it could also be seen as information about others who share the same genetic heritage (e.g. family members). Some of the national surveys indicate that some respondents take the view that any information found through the research must be made available to the original donor of the information.[28] For others, surveys find that the right not to know is strong: the right to remain in ignorance is important, whether you are the donor of the original information or not.[29] Some surveys find that some see that there is a moral dimension to receiving the information.[30] For others, some surveys indicate that receiving the information is about the knowledge that it is not just for them but for family members.[31]

---

26 For example, in Broadata (2001) 65 per cent of the respondents to the UK-based *Whose Hands on Your Genes?* public consultation felt that genetic information should not be placed on a medical research database without consent, while 11 per cent thought it should, and 20 per cent said 'it depends'.

27 See Cragg Ross Dawson (2000) where one opinion from UK focus groups was that surgical waste is unhealthy material and consent for its use would not always be a requirement.

28 For example, Cragg Ross Dawson (2000) report that the general feeling from UK focus groups was that access to any information found about an individual was their right. In Norway, Skolbekken et al. (2005) found that the respondents (linked to a Norwegian genetic research project) agreed that feedback from that research would be good if it could help to prevent future disease. This response was also found by Hoeyer et al. (2004) who asked members of the Swedish general public, the majority of whom agreed as long as the data was certain and treatment would be available.

29 For example, Cousins et al. (2005) found that 18 per cent of the general public in the Republic of Ireland would not wish to know results from the research if there was no prevention or treatment for the disease disclosed, and 8 per cent would not wish to be told even if the disease was preventable or curable. Cragg Ross Dawson (2000) found that some members of UK focus groups felt they only wanted to know research results if the condition was treatable, and others did not want to know at all about anything found through the research.

30 See Haddow et al. (2004: 13) who reported that in their UK focus groups it was suggested that 'there was a moral imperative to make someone aware they were at risk of a genetic condition'.

31 See for example Richards et al. (2003) where women reported that they opted to receive information about epidemiological breast cancer screening for a variety of reasons

The issue of the right to know and the right not to know brings the shared nature of genetic information into sharp focus. The disclosure by one person in, for example, a family or other genetic group may necessarily inform another person in that group of their own genetic status, or will at least alert another person to their potential genetic risk. This, however, places the rights of the second person to a very large degree in the hands of the first. To an extent this mirrors that tension between Article 8 and Article 10 of the European Convention on Human Rights (1950): the right to private and family life is in tension with the right to freedom of expression, and the outcome is a matter of often painful judgement. In surveys, respondents indicate that they would tell other family members about results that they had received,[32] or should tell blood relatives about such results.[33] An Austrian citizen's conference concluded that this decision, even if the disease is severe, should be left in the hands of the patient – i.e. the person being tested (Austrian Council for Research and Technology Development 2003). The other side of the question has been, and still needs to be asked: in what circumstances does the broader genetic group's right to know override the interests of privacy of the tested person?[34]

There is then a sense of a desire for some element of control over the use of the data, which may be similar to ownership rights, in the reported opinions expressed by some survey participants. At the same time, each survey reports a percentage of respondents who for whatever reason hold the opposite view, sometimes that they do not see the loss of control as a problem, or that they are willing to give up their rights for a 'social good' such as genetic research, although the extent to which this is true is also not very well explored in the surveys. The implications for 'group rights' in relation to control over the data are also shown to be not fully explored in the surveys. The following section discusses the possibilities for incorporating the expressed interests into a regulatory regime.

Again, we stress that the survey information reported above is given to show the range of responses to a variety of similar questions in surveys. The surveys

---

including because of concerns of inherited breast cancer in the family. However, the same study also indicated that some women opted not to receive the information because they were old with grandchildren or young with young children. Some who had received results found it difficult to tell family members because of the weight of the information.

32 For example, in Morren et al. (2007) 70.9 per cent and 65.8 per cent of respondents from a postal survey of chronic disease sufferers in the Netherlands indicated that they would inform their children and siblings respectively about the results of genetic tests.

33 For example, an international study by Wertz et al. (2003) on genetic counselling found that 79 per cent of German genetics patients felt that they should disclose the results of genetic tests to their blood relatives, if it were relevant to the relatives' health or reproductive decisions.

34 The public consultation in the UK by the Human Genetics Commission (2001) on *Whose Hands on Your Genes?* asked respondents to consider when non-consensual disclosure may be allowable in a family setting. See the responses online at: http://www.hgc.gov.uk/client/Content.asp?ContentId=387 [accessed 4 April 2009].

have differences in their design and this, of course, can lead to difficulties in making comparisons between the surveys. However, what is clear is that a variety of opinions about expectations of privacy in relation to genetic information and biobanking are expressed by participants when they are surveyed. There is a picture emerging in relation to this new field, but our review suggests that it is not yet complete enough to be able to say that there is a body of data which enables one to say, for the purposes of regulation particularly, that this is a complete or sufficient map of the European peoples' detailed perspectives on this issue. However, what is there indicates that this is possible and should be undertaken. Further, the responses in each of the individual studies invite a different analysis: the survey responses show (at least implicitly) a range of different concepts of privacy that underpin respondents' answers specifically about genetic information and biobanking – how do the different perspectives expressed by participants in surveys relate to the broad range of more theoretical (European) concepts of privacy?

Expressed Interests and Concepts of Privacy

The previous question depends on the more fundamental question: what is privacy? PRIVILEGED does not seek to create a new concept of privacy, rather asks what the range of concepts of privacy is. This is, again, with a view to providing the regulators with sufficient information about the choices available to encourage informed legislating. Therefore, alongside the literature review of the surveys of citizens and their opinions about the interests that may be engaged by research using genetic information and biobanking, PRIVILEGED has also undertaken a literature review of work discussing the nature of privacy. These show a wide variety of theoretical perspectives, in many cases grounded in empirical work on perceptions of privacy in different situations. Our conclusion about these concepts is that privacy remains a genuinely contested concept with some material differences between alternatives. Shifting the conceptual framework brings different interests expressed by the public into, and out of, the privacy frame. The concept of privacy employed by the law is open to development. In this way, determining the concept of privacy employed by the law is an important part of determining the extent of the protection it offers. Westin famously expressed privacy as:

> the claim of individuals, groups, or institutions to determine for themselves when, how, and to what extent information about them is communicated to others. Viewed in terms of the relation of the individual to social participation, privacy is the voluntary and temporary withdrawal of a person from the general society through physical or psychological means, either in a state of solitude or small group intimacy or, when among large groups, in a condition of anonymity or reserve. (Westin 1967)

In 1975, Irwin Altman emphasized the issue of *access* within a rather simpler definition of privacy as the 'selective control of access to the self or one's group' (Altman 1975). This itself can be compared with an attempt in 1977 to provide a consensus definition of privacy by Stephen Margulis who proposed that

> [p]rivacy, as a whole or in part, represents control over transactions between person(s) and other(s), the ultimate aim of which is to enhance autonomy and/or to minimize vulnerability. (Margulis 2003)

Within the different ideas of privacy expressed by just these three authors we can see common themes of 'access' and 'control' but also see them expressed with differing emphasis (e.g. between environment and information; between cultural contingency and universality). When seeking to elucidate the different dimensions of privacy evident within more recent academic writing Anita Allen distinguished between four different dimensions of privacy: 'informational, decisional, physical and proprietary' (1997: 31):

> (1) informational privacy concerns about access to personal information; (2) physical privacy concerns about access to persons and personal spaces; (3) decisional privacy concerns about governmental and other third-party interference with personal choices; and (4) proprietary privacy concerns about the appropriation and ownership of interests in human personality. (Allen 1997: 33)

Again, within each of these dimensions the ideas of 'access' and 'control' are evident, but, less obvious is the idea that privacy is something that may relate to functions and processes between *groups* as well as individuals; an idea much clearer in the concepts of privacy advanced by Westin and Altman. Writing in 2003, Margulis noted also that his earlier (1977) definition of privacy 'failed to note that, in the privacy literature, control over transactions usually entailed limits on or regulation of access to self (Allen, 1988), sometimes to groups (e.g. Altman, 1975), and occasionally to larger collectives such as organizations (e.g. Westin, 1967)' (Margulis 2003: 245).

There are then considerable overlaps between different concepts of privacy within the literature, but the definitions and concepts of privacy that can be found are far from synonymous. Our reading of the literature would in fact suggest that there are (at least) three different *kinds* of potentially significant variation that can be observed: what we are tentatively labelling 'normative', 'transactional' and 'relational' concepts of privacy. We will be focusing here on just one of these; that most relevant to the possibility of 'collective interests' in privacy; and, which we call the 'relational variable'.[35]

35 We are currently, and tentatively, labelling three kinds of variation between different concepts of privacy: (1) 'normative'; (2) 'transactional'; and (3) 'relational'. The 'normative variable' relates to the normative basis of the privacy concept: the underlying justification for

The 'relational variable' seeks to capture the range of opinion on the size of the social unit (from a single individual upwards) capable of being the subject or object of privacy claims. We anticipate that the possibility of 'group privacy' will become particularly important when we later come to compare the existing legal framework with the range of possibilities. For now, however, we wish simply to emphasize the possibility of holding a concept of privacy capable of accommodating the idea of group privacy and to locate some of those interests expressed in research using genetic databases and biobanks alongside such a concept.

'Group' interests in privacy

Bennett and Raab contend that:

> The [current] privacy paradigm rests on a conception of society as comprising relatively autonomous *individuals*. … The modern claim to privacy, then, is based on a notion of a boundary, between the individual and other individuals, and between the individual and the state. (Bennett and Raab 2006: 4)

We would suggest, although we will not argue the point here, that this current paradigm can be clearly seen reflected within the law's understanding of privacy (see Taylor 2006 and 2007). In discussion of research using genetic data, there is often a tension painted between the individual interest in controlling 'their data' and the 'public interest' in the use of that data (see, for example, UK Academy of Medical Sciences 2006). We would suggest that the characterization of that conflict as one between 'privacy' and 'public interest' adopts a particularly individualistic concept of privacy as its foundation. Some of the interests that individuals have expressed in their genetic data might clearly be described as interests in 'other people's data'. They may be interests in data held about family members or data processed about groups with which individuals are associated. They *might* also be described as 'privacy interests'.

When Rabbi Moshe Tendler, objected to the focus on Ashkenazi Jews in genetic research, he is reported to have said:

> Why are you focusing only on the Ashkenazi Jews, giving the world the impression that we have all the bad genes? … [Others] may not know my name, but I've been identified as a member of the Ashkenazi Jewish community who carries bad genes. They had no right to do that to me. (Davis 2000: 39)

---

claims to privacy (see Laurie 2002). As well as recognizing potential variation with respect to the underlying normative basis of a concept, the 'content' of claims to privacy may also manifest differently in different social circumstances. The varying 'content' of potential privacy claims is described by the (trans)actional variable.

An interest in the research that is conducted upon the group with which one is associated would be much more readily accommodated as a 'privacy interest' by some concepts of privacy than it would others (see, for example, Alpert 2003).

Expressed Interests, Legitimacy and Challenges for Regulation

The PRIVILEGED literature surveys have identified, as indicated above, that there is a wide range of opinions about the interests that are engaged by research using genetic databases and biobanks. Is this a surprising range of expressed opinions? Whereas it is not possible in the data to directly attribute the holding of particular positions to particular beliefs or socio-political views (because the surveys we have seen do not ask such questions directly), the responses resonate with the usual range of European sensitivities as they are expressed in other debates. For example, some responses about the morality of certain biotechnological uses of the genetic information could sit with religious views taken in relation to embryos, some of the views about the commercial ownership of genetic information sit well with liberalism and free market ideologies, where as some of the responses indicating duties to consider the public interest above one's own have strong collectivist overtones. However, these debates and views are not particularly 'European'.

What is fundamental to ask is how should the range of answers described for each question in each survey be read? Whilst there is in most cases a majority position, each majority discloses a minority position, or range of minority positions, which represent the sensitivities of individuals. Many of these minority positions are well inside the range of sensitivities that are accepted in Europe as legitimate political, philosophical, social and religious beliefs. How far can these divergent responses be accommodated in the regulation of privacy in genetic information used in research and in biobanks, and how far should the full range of interests expressed by the citizens be accommodated in regulations of the area? Thus, some express a view that genetic data and biobanks should not give rise to corporate property whereas others see this as necessary or desirable; some see the ownership of genetic data by the State as dangerous, others see it as appropriate. Likewise, in the reaction to the rights of individuals within groups, a wide range of socially acceptable positions is reflected. In these circumstances, for a regulator to seek to simply reflect 'the majority view' is not enough if the genetic project is to be realized fully. Simply taking the majority view will exclude and alienate many who hold other, socially legitimate opinions (i.e. opinions which are established within general mainstream societal values). Whereas a more inclusive regulatory framework, reflecting the broad range of legitimately held opinions, would encourage the participation of such individuals, the narrow regulatory approach alienates them and may cause them to opt out of participation in the genetic project. These may be individuals in minorities defined by their genetic experience; people who would be interesting to include in the genetic project for its purposes, but more importantly whose vulnerability and need for the potential benefits of the project make it important to include.

By way of an example, how would this approach work in relation to one area of opinion expressed in the literatures? Given that some wish to be engaged in research using genetic databases and biobanks only with a full informed consent regime, others wish to be engaged with a broad consent, others do not wish to participate at all, and yet others wish to participate without any consent, and others feel differently about different types of data and tissues, the first challenge to regulators is to determine that range of opinions that *are* to be accommodated within a regulatory framework. Is it acceptable to 'abandon' one's genetic information to a research biobank, or should one remain responsible for determining its use? This is the first and major political decision which requires, at least as a starting point, an understanding of the range of opinions (both theoretical and empirical) in play.

The next challenge is to determine *how* simultaneously to accommodate diverse opinion within the regulation of the area. To the extent that any attempt to accommodate the complete range of expressed interests would frustrate the project of genetic research for public benefit, any insistence that they must be accommodated could also be characterized as unacceptable. However, we would suggest that an iterative process can be imagined that would, if undertaken, progressively identify the widest range of options for participants consistent with effective research. While this might not ultimately settle upon a system of regulation that is most effective for researchers (in the short term), we return to our initial remarks that the nature of research using biobanking and genetic information is such that maximum inclusion is the most desirable position. Regulation should therefore strive to accommodate as many positions as possible by developing individuals' and groups' trust and confidence in the safeguarding of their interests. This does, however, involve both knowing what the expressed interests are and also, operating with legal concepts capable of accommodating them. The maintenance of public trust is clearly a matter of concern to all members of the public, including researchers themselves.

Stage two of the PRIVILEGED project will determine just how far a mismatch currently exists between expressed interests and regulation, at the national and European level. It should be recognized that to react nationally to a 'local range' of expressed interests is not a realistic option for members of the European Union. The context within which these interests are regulated is distinctly European. On the one hand, it is regulated within the context of the European Convention on Human Rights and the tradition of the balance of those rights between the individual and the public interest. On the other, the European Union seeks economic, commercial-driven harmonization, with an increasing social harmonization to underscore this economic landscape.[36] If these three elements – the public's expressed interests, the

36 See, for example the Treaty of Maastricht (1992) with the inclusion of greater social rights alongside the purely economic harmonization of the European Member

human rights tradition, and the community harmonization traditions – are brought together in order to achieve the maximum inclusion of expressed interests whilst holding onto legislative coherence, then this would contribute to the international debate with a European voice that speaks of inclusion and respect.

Where a truly valuable contribution could be made would be in bringing this inclusion and respect to bear on the rights of the individual data subject in relation to the rights of the broader genetic group or groups within which he or she is located. One thing does, however, already appear to be clear. There is sufficient diversity of opinion apparent within existing public attitude surveys, and lack of exploration of public attitudes across certain parts of Europe, that harmonizing research protocols across Europe on the basis of an assumed 'European Perspective' may currently be both a premature and risky endeavour. The perspective taken by the Law in Europe ought to take account of the range of European opinion. There is an opportunity to revisit the legal understanding of privacy to ensure that it is capable of acknowledging the optimal range of interests for effective regulation. An appropriately informed, and accommodating, concept of privacy underlying the work of the European Court of Human Rights, might make significant advance toward ensuring public trust in the important research to be conducted in this area.

Acknowledgements

We are very grateful to Dr Klaus Hoeyer (University of Copenhagen and member of the PRIVILEGED project), and Professor Klasien Horstman and Dr Ine van Hoyweghen (Maastricht University) for their help and comments on this chapter.

Bibliography

Cases

R (S and Marper) v. *Chief Constable of the South Yorkshire Police* [2004] UKHL 39, [2004] 1 W.L.R. 2196, HL, United Kingdom.

Laws

EC Directive 95/46/EC of the European Parliament and of the Council of 24 October 1995 on the protection of individuals with regard to the processing of

States; the Charter of Fundamental Rights of the European Union (2000), the Draft Treaty Establishing a Constitution for Europe (2004), and the Treaty of Lisbon (2007) (although see comments above in footnote 6).

personal data and on the free movement of such data 1995. *Official Journal* L 281, 23/11/1995, Luxembourg: EUR-OP.

Charter of Fundamental Rights of the European Union 2000. *Official Journal* C 364/01 18/12/2000, Luxembourg: EUR-OP. Available at: http://www.europarl.europa.eu/charter/pdf/text_en.pdf [accessed 4 April 2009].

Draft Treaty Establishing a Charter for Europe 2004. *Official Journal* C310, 16 /12/2004, Luxembourg: EUR-OP. Available at: http://eur-lex.europa.eu/en/treaties/dat/12004V/htm/12004V.html [accessed 4 April 2009].

European Convention on Human Rights. 1950 (European Convention for the Protection of Human Rights and Fundamental Freedoms). ETS No. 005, Rome: Council of Europe. Available at: http://conventions.coe.int/Treaty/en/Treaties/Html/005.htm [accessed 4 April 2009].

Treaty of Lisbon. 2007. *Official Journal* C306 17/12/2007, Luxembourg: EUR-OP. Available at: http://eur-lex.europa.eu/JOHtml.do?uri=OJ:C:2007:306:SOM:en:HTML [accessed 4 April 2009].

Treaty of Maastricht. 1992. *Official Journal* C191 29/07/1992, Luxembourg: EUR-OP. Available at: http://eur-lex.europa.eu/en/treaties/dat/11992M/htm/11992M.html#0001000001 [accessed 4 April 2009].

References

The Academy of Medical Sciences (UK). 2006. *Personal Data for Public Good: Using Health Information in Medical Research.* [Online] Available at: http://www.acmedsci.ac.uk/p48prid5.html [accessed 4 April 2009].

Allen, A.L. 1988. *Uneasy Access: Privacy for Women in a Free Society.* New York: Rowman & Littlefield Publishers Inc.

Allen, A.L. 1997. Genetic Privacy: Emerging Concepts and Values, in *Genetic Secrets: Protecting Privacy and Confidentiality in the Genetic Era*, edited by M.A. Rothstein. Connecticut: Yale University Press, 31-60.

Alpert, S.A. 2003. Protecting Medical Privacy: Challenges in the Age of Genetic Information. *Journal of Social Issues*, 59(2), 301-322.

Altman, I. 1975. *The Environment and Social Behaviour: Privacy, Personal Space, Territory, Crowding*. Monterey: Brooks/Cole Publishing Company.

Austrian Council for Research and Technology Development 2003. *Genetic data: Where From, Where To, What For? (Citizen's conference, Vienna, 20-23 June 2003*). Information available at: http://science.orf.at/science/news/79025 [accessed 4 April 2009].

Bennett, C.J. and Raab, C.D. 2006. *The Governance of Privacy: Policy Instruments in Global Perspective*. Cambridge: MIT Press.

Broadata. 2001. *Whose Hands on Your Genes? Analysis of Responses to the Tick Box Summary Section* [Online: UK Human Genetics Commission]. Available at: http://www.hgc.gov.uk/UploadDocs/Contents/Documents/iievidence_whoygsummary_responses.pdf [accessed 4 April 2009].

Calnan, M., Montaner, D. and Horne R. 2005. How Acceptable are Innovative Health-care Technologies? A Survey of Public Beliefs and Attitudes in England and Wales. *Social Science & Medicine*, 60, 1937-1948.

Cousins, G., McGee, H., Ring, L., Conroy, R., Kay, E., Croke, D. and Tomkin, D. 2005. *Public Perceptions of Biomedical Research: A Survey of the General Population in Ireland.* Dublin: Health Research Board.

Cragg Ross Dawson. 2000. *Public Perceptions of the Collection and Use of Human Biological Samples (prepared for the Wellcome Trust and the Medical Research Council).* London: The Wellcome Trust.

Davis, D.S. 2000. Groups, Communities, and Contested Identities in Genetic Research. *Hastings Centre Reports*, 30(6), 38-45.

Eriksson, K.E. 2007. Sweden, in *The Ethics and Governance of Human Genetic Databases: European Perspectives*, edited by M. Hayry, R. Chadwick, V. Árnason and G. Árnason. Cambridge: Cambridge University Press.

European Commission. 2005. *Social Values, Science and Technology. Special Eurobarometer 225/Wave 63.1.* [Online] Available at: http://ec.europa.eu/public_opinion/archives/ebs/ebs_225_report_en.pdf [accessed 4 April 2009].

Gaskell, G. et al. 2006 *Europeans and Biotechnology in 2005: Patterns and Trends. Final report on Eurobarometer 64.3* [Online] Available at: http://ec.europa.eu/research/biosociety/pdf/eb_64_3_final_report_second_edition_july_06.pdf [accessed 4 April 2009].

Gudmundsdóttir, M.L. and Nordal S. 2007. Iceland, in *The Ethics and Governance of Human Genetic Databases: European Perspectives*, edited by M. Hayry, R. Chadwick, V. Árnason and G. Árnason. Cambridge: Cambridge University Press.

Haddow, G., Cunningham-Burley, S., Bruce, A. and Parry, S. 2004. *Generation Scotland Preliminary Consultation Exercise 2003-2004: Public Stakeholder Views from Focus Groups and Interviews.* Edinburgh: ESRC Innogen Centre, University of Edinburgh.

Hoeyer, K. 2002. Conflicting Notions of Personhood in Genetic Research. *Anthropology Today*, 18(5), 9-13.

Hoeyer, K., Olofsson, B.-O., Mörndal, T. and Lynöe, N. 2004. Informed Consent and Biobanks: A Population-Based Study of Attitudes Towards Tissue Donation for Genetic Research. *Scandinavian Journal of Public Health*, 32(3), 224-229.

Hoeyer, K., Olofsson, B.-O., Mörndal, T. and Lynöe, N. 2005. The Ethics of Research Using Biobanks: Reason to Question the Importance Attributed Informed Consent. *Archives of Internal Medicine*, 165, 97-100.

Human Genetics Commission. 2001. *Whose Hands on your Genes? Consultation Responses.* [Online] Available at: http://www.hgc.gov.uk/client/Content.asp?ContentId=387 [accessed 4 April 2009].

Human Genetics Commission. 2002. *Inside Information: Balancing Interests in the Use of Personal Genetic Data. A Summary Report.* [Online, May 2002].

Available at: http://www.hgc.gov.uk/UploadDocs/DocPub/Document/inside information_summary.pdf [accessed 4 April 2009].

King Baudouin Foundation. 2003. *Is it in My Genes? Final Report (Lire dans mes génes? Avis des citoyens) (Citizen's Conference, January to March 2003, Belgium)*. Available at: http://www.kbs-frb.be/publication.aspx?id=178108&LangType=2060 [accessed 4 April 2009].

Korts, P. 2004. Becoming Masters of Our Genes: Public Acceptance of the Estonian Genome Project, in *Blood and Data. Ethical, Legal and Social Aspects of Human Genetic Databases*, edited by G. Àrnason, S. Nordal and V. Àrnason. Reykjavik: University of Iceland Press, 189-193.

Laurie, G. 2002. *Genetic Privacy: A Challenge to Medico-Legal Norms*. Cambridge: Cambridge University Press.

Levitt, M. and Weldon, S. 2005. A Well Placed Trust?: Public Perceptions of the Governance of DNA Databases. *Critical Public Health*, 15(4), 311-321.

Lind, U., Mose T. and Knudsen L.E. 2007. Participation in Environmental Health Research by Placenta Donation – A Perception Study. *Environmental Health*, 22(6), 36.

Margulis, S. 2003. Privacy as a Social Issue and a Behavioural Concept. *Journal of Social Issues*, 59(2), 243-261.

MORI Social Research. 2001. *Major Attitudes to Human Genetics: People's Panel Quantitative Survey Conducted for the Human Genetics Commission*. London: Human Genetics Commission.

Morren, M., Rijken, M., Baanders, A.N. and Bensing, J. 2007. Perceived Genetic Knowledge, Attitudes Towards Genetic Testing, and the Relationship between these Among Patients with a Chronic Disease. *Patient Education and Counseling*, 65, 197-204.

Pin, R. and Gutteling, J.M. 2005. *Publieksonderzoek Genomics 2005 (Public Survey Genomics)*. Enschede: Universiteit Twente.

Rand, A. 1943. *The Fountainhead*. London: Signet Books.

Richards, M.P.M., Ponder, M., Pharoah, P., Everest, S. and Mackay, J. 2003. Issues of Consent and Feedback in a Genetic Epidemiological Study of Women with Breast Cancer. *Journal of Medical Ethics*, 29, 93-96.

SINICA. 2006. *Taiwan Genomic Survey* (sponsored by the National Science Council of Taiwan).

Skolbekken, J-A., Ursin, L.Ø., Solberg, B., Christensen, E. and Ytterhus, B. 2005. Not Worth the Paper it's Written On? Informed Consent and Biobank Research in a Norwegian Context. *Critical Public Health*, 15(4), 335-347.

Stichting Consument en Biotechnologie. 2002. *Publieksonderzoek (Public Survey) Genomics*. The Hague: Stichting Consument en Biotechnologie.

Taylor, M.J. 2006. Data Protection, Shared (Genetic) Data and Genetic Discrimination. *Medical Law International*, 8(1), 51-77.

Taylor, M.J. 2007. Regulating Personal Data in a Shared World. *Personalised Medicine*, 4(4), 471-477.

Tupasela, A., Snell, K., Sihvo S., Jallinoja, P., Aro, A.R. and Hemminki, E. 2007. Väestön suhtautuminen lääketieteellisten näytekokoelmien uudelleen käyttöön ja biopankkeihin [The Perceptions of Finn's Towards the Biomedical Use of Tissue Collections and Biobanks]. *Suomen lääkärilehti (Finnish Medical Journal)*, 62(51-52), 4780-4782.

Wertz, D., Irmgard, N. and Gerhard, W. 2003. Patient and Professional Responsibilities in Genetic Counselling, in *Patienten-und Bürgerau fklärung über genetische Risikofaktoren*, edited by H.-M. Sass and P. Schröder. Münster: Lit-Verlag.

Westin, A. 1967. *Privacy and Freedom*. New York: Atheneum.

Chapter 10
Feeding back Significant Findings to Participants and Relatives

Loane Skene

The decision whether to feed back findings to research participants has been the subject of much unresolved debate in medical research, as it challenges the boundary between the clinic and the research context (Wolf, Paradise and Caga-anan 2008). Despite this, there has been relatively little discussion of the issue of feedback in the context of biobanks. Feedback is particularly significant for biobanks and other large-scale repositories of human tissue and medical information,[1] where there maybe an on-going relationship with participants over many years and over time, the knowledge about health risks is likely to increase. This issue will become even more acute with the use of new approaches in genomic research, such as genome-wide association studies (GWAS) (Cho 2008), and the future use of whole genome sequencing technology, where the possibility of incidental findings increases (Van Ness 2008, Wolf et al. 2008). This chapter considers the nature, extent and enforceability of the legal duties that custodians of biobanks and the scientists and others who use tissue or information from these sources may owe to participants – and possibly also to their close blood relatives – in the case of significant findings. The nature of those duties will depend ultimately on the precise legal relationship between the parties and the legal principles that may be applied. This chapter focuses principally on Australian law, but this discussion will be important for biobanks based in other countries, as similar legal issues will arise and a similar analysis will need to be undertaken.

1 The Australian Law Reform Commission (ALRC) mentioned a range of 'health information data bases' in Australia: the Medicare, Pharmaceutical Benefits Program databases; the cervical cancer register; the ANZ Dialysis and Transplant Registry; and the Menzies Centre for Population Research which collects genealogical data, genetic samples, and health information supplied by donors, to search for genetic causes of disease (ALRC 2008). These might all be included in the repositories discussed in this chapter but, for ease of reference in this chapter, the various repositories of human tissue and medical information are called 'biobanks'.

'Significant Findings'

Significant findings may arise from the results of the planned research, or incidentally, without having been anticipated. The meaning of 'significant' in the current context is open to interpretation. A clear example of a clinically significant finding is a link between a genetic mutation and a serious medical condition where people who are warned of that link could take measures to protect themselves by tests to determine their susceptibility and treatment, if these are available and necessary. However, other significant findings may not be directly clinically significant. For example, they may have reproductive implications, or be important in other ways, such as non-divulged non-paternity.

Determining whether a finding is 'significant' for legal purposes, that is, giving rise to an ethical or legal duty to provide feedback to the participants, has been the subject of much debate in other contexts. In 2004, the National Heart, Lung and Blood Institute (NHLBI) of the National Institutes of Health, established a working group that recommended that in the case of genetic conditions, feedback should be given if: (1) the risk for disease is significant; (2) the disease has important health implications (i.e. fatal or substantial morbidity); and (3) there is a proven therapeutic or preventative intervention available' (Wolf et al. 2008: 230, Table 3, Van Ness 2008). Johnston and Kaye argue that, in the case of population biobanks, an obligation exists in the case of a 'serious, treatable condition' (Johnston and Kaye 2004). These statements neatly sum up the two main features that make findings ethically or legally significant – the seriousness of the condition and the availability of a medical intervention; though, in Australia, the wider concept 'bears on the wellbeing of participants', discussed below, should also be noted.

Duty of Biobanks – and Others – to Notify Participants

Some biobanks, such as the UK Biobank, have a policy that they will not provide information directly to participants about the results of research, though general information may be published on the web or in newsletters. This information may be available to the world at large, or limited to participants in the research. Indeed, a biobank might argue that participants are told at the outset that they will not be given personal information about research results, that they agree to participate on that basis and that there is therefore no legal obligation to inform them of the results. But it is open to question whether this practice is legally supportable.

Legal Relationship between Biobanks and Participants

There are a number of legal regimes that may govern the relationship between biobanks and participants. The first is the statute establishing the biobank, if it operates under a statute. The second is a contract between the participants and the biobank. The third is the law of tort, under which a duty of care may be owed to research participants (and possibly relatives) by the biobanks and the researchers who use the resources. Finally, property law may also be argued to be relevant, though that argument has not been accepted where it has been advanced in the few cases that have come before the courts (Wolf, Paradise and Caga-anan 2008: 372-3, in which the authors discuss the cases *Greenberg v. Miami Children's Hospital Research Institute* (264 F. Supp. 2d 1064 (S.D. Fla. 2003)); and *Ande v. Rock* (647 N.W. 2d, 265 (Wis. App. 2002)) which they note 'claim[ed] a property right not only to biologic samples, but also to the results of research using those samples' and 'have been largely unsuccessful').

Statutory provisions

Some biobanks operate under a statute, such as tumour registers. For example, hospitals and pathology services in Victoria are authorized by the *Cancer Act 1958* (Vic) s.60(1) to report the results of tests for cancer to the Cancer Council or a cancer register.[2] The Schedules to the *Cancer (Reporting) Regulations 2002* (Vic) list matters to be reported,[3] some of which may be significant if a link is established between particular observations and the incidence or prognosis of a disease. A person whose details have been reported to a cancer register could presumably gain access to that information under the *Freedom of Information Act 1982* (Vic) s 13, but the obligation of confidentiality with regard to cancer test results imposed by the *Cancer Act 1958* (Vic) ss.61 and 62(6) would prevent information from the register being disclosed to other people, except for research approved by an ethics committee under guidelines of the National Health and Medical Research Council (NHMRC); or to clarify the accuracy of information with the Cancer Council: ss 61A, 61B.

There are two provisions in the Act that might possibly impose an obligation to 'follow up' a significant finding. Under s.62(5) the functions of 'an organisation that maintains a prescribed register' are '(a) to follow up positive results from cancer tests; and (b) to send reminder notices when persons whose names appear in the register are due for cancer tests'. Under s.61(3), information provided by employees of the Cancer Council for approved research is subject to NHMRC guidelines, which include superseding NHMRC guidelines that cover the same subject matter: s.61A(3).

---

2 Reporting is not mandated as the person concerned may object: *Cancer Act 1958* (Vic) s.62(3).

3 Matters to be reported by hospitals include: patient details; investigations relevant to diagnosis of cancer; laterality, morphology and grade/differentiation of primary cancer; staging details (if available): degree of spread of cancer, localized to the tissue of origin, invasion of adjacent tissue or organs, regional lymph nodes, distant metastases, and the like: R 5, Sch 2.

The latest version of the NHMRC guidelines was published in 2007: the *National Statement on Ethical Conduct in Human Research* (NHMRC 2007). Chapter 3.2 includes specific provisions regarding databanks. Para 3.2.6 states:

> Whenever research using re-identifiable data[4] reveals information that bears on the wellbeing of participants, researchers have an *obligation to consider how to make that information available to the participants*. Where individual notification is warranted, the custodian of the data will need to take all reasonable steps to re-identify those data. (emphasis added)

This statement is in stronger terms, both in the obligation to provide feedback and in the circumstances in which it must be considered[5] – than para 14.13 in the earlier *National Statement on Ethical Conduct in Human Research* (NHMRC 1999), dealing with 'epidemiological research':

> 14.13 If in the course of epidemiological research new knowledge of clinical relevance is obtained, or existing treatment is thought to need alteration, that knowledge should be disclosed to the appropriate health authorities and, *wherever possible*, participants and their usual medical attendants should be informed. (emphasis added)

Because these guidelines are incorporated by reference in the statute, they may be regarded as having the force of the statute and so impose 'an obligation to consider how to make that information available to the participants' including possible 'individual notification'. Thus the obligation imposed on employees of the Cancer Council relating to providing information 'that bears on the wellbeing of participants' are to *consider* how to make the information available, not to actually make it available, or to make it available in a particular way.

Contract

The legal relationship between a biobank and participants may be contractual.[6] If that is the case, the terms of the contract will generally be implied and evidenced by the consent form signed by the participant and other information provided to

---

4 The NHMRC *Statement* says that 'human tissue samples should always be regarded as, in principle, re-identifiable': Ch 3.2.

5 Compare the words 'information that bears on the wellbeing of participants' with the features of a 'significant finding' in the text above (headed 'Significant Findings'), especially the availability of a clinical intervention.

6 Graeme Laurie suggested in his paper at the Oxford Conference on Biobanks that 'benefit sharing is a better model to motivate and retain participants (and to inspire public confidence) to avoid the legal machinations of property or contract models': 'Governing Biobanks – What are the challenges?' St Anne's College, Oxford, UK, 24-26 June 2008.

the participant at the outset, such as a plain language form and information on the biobank's website.

Under the contract, participants will authorize the use of their tissue or information for the purpose of the research described in the form – that is, for research in general, or as stated in the form. Note, with regard to the use of tissue, that this arrangement is an *authority* to use the tissue for the stated purposes. It is not a gift or an assignment of the tissue. That would imply that the participant had a proprietary interest in the tissue, which is unlikely to be the case.

Some commentators have argued that the law should recognize a right of 'ownership' or ongoing control in relation to excised human tissue (Laurie 2002: 321, Morgan 2001: 83-104) but that has not been accepted as a general legal principle. Until recently, the legal position was that people do not have any proprietary interests in their excised tissue. A limited exception was recognized in the recent decision of *Yearworth* in England, in which the Court of Appeal held that the appellants, who had deposited semen samples for freezing before they undertook treatment for cancer, 'for the purposes of a claim in negligence … had ownership of the sperm which they had ejaculated' (*Yearworth* 2009: [45] para f). The reason for this decision was that, when the sperm was deposited, 'the sole object … was that, in certain events, it might later be used for their benefit' (*Jonathan Yearworth and Others v. North Bristol NHS Trust* [2009] EWCA Civ 37: [45] para f). However, there is no suggestion that the participants who deposit their tissue in a biobank have any expectation that it will be available to them later for their own use, so the *Yearworth* argument seems not to be relevant to tissue deposited in biobanks, or to information derived from it; and property arguments have not been accepted by courts in other cases (Wolf et al. 2008).

In *Yearworth*, the Court of Appeal also held that the unit in which the sperm had been deposited was liable in bailment, as a gratuitous bailee (at [51]); that the arrangements were 'closely akin to contracts' (at [57]); and that 'modest recovery' could be awarded for 'mental distress' against a gratuitous bailee in such circumstances as well as damages in tort (at [59]). However, like ownership, bailment is based on ongoing control which is absent when tissue is donated with the intention that it will not be available to the participants for their later use, and also that it may be used in research. A biobank is more like the university in the American case, *Washington University v. Catalona* (*Washington University v. Catalona* (2006) 437 F Supp 2d 985, US District Court, Eastern District of Missouri)[7] which, as the Court of Appeal said in *Yearworth* was a donee and not a bailee because 'the donors had abandoned any possessory interest in the tissue' (at [48]). For this reason, researchers who have used tissue to develop a new product,

---

However, although this seems a noble and desirable model, it is difficult to see how one can prevent a contract or other legal relationship arising.

7 The decision was later affirmed by the US Court of Appeals, 8th Circuit: 490 F 3d 667.

such as a diagnostic test or cell line, may gain a proprietary interest in it (Skene 2002a, Skene 2002b).

If the relationship between a biobank and the research participants is contractual, it is conceivable that they could state specific terms in the contract and that those terms would then be enforceable because of the contract between them. The biobank could specify that it will not contact participants, even if there are significant findings, or that it will notify participants only by posting significant findings on its web site. But, even if such an 'exemption clause' was permitted,[8] it would be interpreted by the court, like all exemption clauses, *against* the party seeking to rely on it, such as the biobank or third party. If there was any ambiguity about the terms of the contract they would be unwise to rely on such a term.

If there is no specific term in the contract between the biobank and participants requiring participants to be notified of significant findings, it is possible that such a term might be implied. For example, if the implied terms in the contract include a requirement of compliance with professional guidelines, those guidelines may recommend disclosure in certain circumstances, such as Chapter 3.2 of the *National Statement* (NHMRC 2007), quoted above. Also, the research ethics committee overseeing the biobank may impose an obligation to comply with the *Statement* as a condition of funding and the participant might argue that this means that it should be implied in the contract between the participant and the biobank. If the biobank is privately funded, it may consider itself free of obligations arising under the *Statement*, because its funding is not derived from a public source. However, the terms of the *Statement* may still have legal force as evidence of accepted practice or 'reasonable care' and therefore be implied in the contract even if the biobank is privately funded. But even if such a term was implied, it would be subject to interpretation as suggested earlier, so that the biobank's obligation would be to '*consider* how to make … information available to the participants' (emphasis added).

Alternatively, participants might require the biobank to give an undertaking in the contract that it *will* contact them about significant findings that come to its attention. The biobank would then need to include a similar term in the contracts

8 Wolf et al. (2008) state that 'Researchers should not be able to use the informed consent process to 'contract out' of their ethical obligations … [any] more than they could ask individuals to participate in research with an unreasonable balance of risks and benefits' (p. 375); that 'the regulations forbid exculpatory language in the informed consent process …' (p. 367); and that exculpatory language may not protect the researcher from liability for failure to disclose if an incidental finding 'is of clear health significance to the research participant' (p. 375). However, most of the research that they discuss involves a clear relationship between researchers and participants and large scale holdings like biobanks are barely mentioned. Biobanks may be subject to different considerations, as two US regulatory bodies have 'taken the position that research on archived data that has been anonymized is exempt from regulation by the Common Rule' (p. 364).

it enters into with third parties and require them to have a similar term in any contracts that they enter into with others.

Thus, in summary, if the relationship is contractual, the rights and duties of the parties would be governed by the contract. In the absence of an express term dealing with feedback, the contract is likely to be construed so as to include at most the obligation in para 3.2.6 of the *Statement* to 'consider how to make … information available to the participants'. If a biobank fails to make information available, then participants may have a cause of action against the biobank for breach of contact (and possibly also in tort, as suggested below) but the contract will be open to interpretation. If the significant finding is made by a third party, such as a researcher who uses the biobank, that person could have a similar contractual obligation to notify participants under their contract with the biobank. However, they could be specifically exempted from doing so by their contract with the biobank, though that exemption might be open to challenge, as suggested above. If the researcher does have a duty to make information available and does not do so, the biobank may be joined in the litigation if the researcher is sued by a participant for not making information available about a potential risk.

Tort: Duty to warn?

Whether the relationship between a biobank and the participants is contractual or not, rights and duties may also arise in tort. These may exist independently or they may be in addition to rights and duties in contract, if there is a contract between the parties.

The biobank and possibly third parties who acquire tissue or information for research would seem to be sufficiently proximate to the participants to meet Lord Atkin's 'neighbour' test in *Donoghue v. Stevenson* (*Donoghue v. Stevenson* [1932] AC 56 (HL)) so as to give rise to a duty to take reasonable care not to injure the participants.[9] This is more arguable in cases where the researcher is also clinically treating the participant. In Australia, it is clear that clinicians owe a duty of care to any patients they are treating but no courts have yet considered the possible duty of care owed by researchers who are not in a clinical relationship with participants. In the United Kingdom, on the other hand, it was established in the Creutzfeldt-Jakob litigation that researchers do owe a duty of care to research participants and that this is 'akin to that of doctor and patient, one of close proximity' (*Creutzfeldt-Jakob Disease Litigation* (1996) QB 54 BMLR 8: 9) However, the duty in that case was found in a clinical trial which then became a general therapeutic programme. It was largely because it was a therapeutic programme that a duty of care was established without considering proximity.

9 The duty discussed here is a legal duty. There may, of course, be a moral duty to warn, as Graeme Laurie suggested in his paper at the Oxford Conference: note 6 above. He suggested that the basis of a moral duty might arise from the obligation of 'stewardship' and the promotion of trust, or be a 'payback' based on reciprocity.

In Canada, the law has recognized that researchers owe a duty of care to research participants (*Halushka v. University of Saskatchewan* [1965] 52 W.W.R. 608 (Sask. C.A in 1965), *Weiss v. Solomon* (1989), 48 C.C.L.T. 280 (Que. Sup. Ct). 6), Zimmerman 2005). Similarly, in the United States, some commentators have argued that researchers have a duty to disclose incidental findings 'of potential clinical or reproductive significance' to participants because of the 'special relationship between [them], in which the researcher is privy to information about the participant based on their interaction' (Wolf et al. 2008: 367). No principle of this kind has been recognized in Australian law and in any event, even the principle suggested by those commentators may not extend to findings from the use of large-scale biobanks.[10]

However, even if there is a duty of care, it does not necessarily require the biobank to warn participants of significant findings, and, *a fortiori*, to warn others who may have a similar genetic risk, such as close blood relatives of the participant. Courts in Australia have been reluctant to impose a legal duty to warn and, even in the United States, where such a duty was recognized in the *Tarasoff* case (*Tarasoff v. Regents of the University of California*, 17 Cal. 3d 425, 551 P.2d 334, 131 Cal. Rptr. 14 (Cal. 1976)) the courts have been cautious. The *Tarasoff* principle has not been applied in all states and, in the few cases coming before the courts involving a genetic risk, some judges have said that it is enough for the doctor to advise the patient to pass the information on to relatives. Doctors have generally not been held liable for not warning relatives directly. Thus, in *Pate v. Threlkel* (*Pate v. Threlkel* 661 So 2d (1995) (SC Florida)) the doctor was held to have a duty to inform relatives but that could be discharged by warning the patient who could be expected to pass the information on to the relatives.[11] This case may be compared with *Safer v. Pack* (*Safer v. Pack* 291 N.J. Super 619, 677 A 2d 1188 (1996) (S.C. New Jersey, App. Div.)) in which the court was not satisfied that a doctor's warning to tell relatives would always be enough.

In Australia, courts have not recognized a general duty of health professionals to warn third parties based on the knowledge of a serious risk to those people, though in one New South Wales case, a doctor was held to have a duty of care to the patient's sexual partner, who had never been his patient (*BT v. Oei* [1999] NSWSC 1082). However, the duty of care owed to the partner would have been fulfilled by the doctor providing appropriate care and advice to the patient, as in *Pate v. Threlkel* above; it did not require a warning to be given directly to the patient's partner.

10 Most of the research that Wolf et al. (2008) discuss involves a clear relationship between researchers and participants, and large-scale holdings like biobanks are barely mentioned. Biobanks may be subject to different considerations. As they say, two US regulatory bodies have 'taken the position that research on archived data that has been anonymized is exempt from regulation by the Common Rule' (p. 364).

11 *Pate v. Threlkel* was followed in *Werner v. Varner* 659 So 2d 278 (1995) (S.C. Florida) though that was not a case of genetic risk.

However, even if the law recognized a legal duty to warn in a clinical context, a biobank could argue that the research context is different and should not give rise to a claim in negligence for failure to warn. It could deny the existence of a 'relationship' giving rise to a duty of care and argue that there is no legal duty to provide information to participants about research findings, much less to relatives. Also, even if there is a duty, they may say that they have not breached that duty. In relation to participants (and possibly relatives), the duty is to take reasonable care. That is open to interpretation. One guide is peer practice and in deciding what 'reasonable care' would require, a court could consider the *National Statement* (NHMRC 2007). The biobank could argue that it had complied with the *Statement* because it had 'considered' how to make information available and, on reasonable grounds, it had decided not to do so. Or, as suggested earlier, there may be a contractual exemption – that is, the participants may have agreed at the outset that no information would be provided (though that exemption would not extend to relatives, who would not be parties to the contract).

Also, even if a duty to warn is established, there may be an issue of causation. The biobank might argue that the cause of the 'injury' – the development or worsening of the participant's genetic condition – was in fact the person's genetic constitution and not the biobank's failure to warn, so that an action in negligence would fail on the ground of causation. An argument that participants might have reduced their potential harm if warned of the risk seems to fall in the category of an action for 'loss of a chance' and the High Court of Australia, in a joint judgment in *Gett v. Tabet*, recently rejected the doctrine of the loss of a chance of a better outcome in medical negligence cases (*Gett v. Tabet* [2009] NSWCA 76 (9 April 2009)).[12] However, *Gett v Tabet* is currently under appeal to the High Court.

Breaching Confidentiality

It is possible, of course, that a biobank might *want* to warn participants or their primary health providers, and even the participants' close blood relatives or other people at risk such as spouses and partners,[13] of findings that may be significant for them. In that case, participants may object to their confidential information being disclosed to other people without their consent and blood relatives may object to genetic information (that also relates to them) being revealed to other people, especially if it identifies their family. Genetic information concerning the incidence and patterns of disease and genetic conditions and traits in the family may be significant to insurers, employers, government agencies and the like.

---

12 The court refused to follow *Rufo v. Hosking* [2004] NSWCA 391; (2004) 61 NSWLR 678 (CA) and *Gavalas v. Singh* [2001] VSCA 23; (2001) 3 VR 404 (CA).

13 The risk to spouses and partners is not direct. It is a potential risk to their children which they could take into account in monitoring their child's health or making future reproductive decisions.

Biobanks should obviously be cautious in disclosing any personal information that may identify individuals or families. Disclosing personal information about people without their consent may breach the right that those people have to confidentiality at common law and may be actionable in tort (negligence and breach of confidence) and contract.[14] It may also contravene privacy legislation[15] if applicable, and be the basis of a complaint to a privacy body[16] or to a registration board ('unprofessional conduct').[17]

However, there are many responses to concerns about breaching privacy. Research participants who donate tissue or provide personal information to a biobank are entitled to have their privacy protected but, if they agree to have their tissue or their information used in research, that agreement constitutes an authority to use the information for the stated purposes, so that it can be lawfully shared among the researchers. This principle applies whether the information is genetic (such as a DNA test), phenotypic, other lifestyle information or any other personal information. For example, under the *Privacy Act 1988* (Cth) Sch 3, NPP 2, an 'organization' (which includes individuals: s 6C) may lawfully 'use or disclose personal information … with the *consent* of the person concerned' (Privacy Act Sch 3, NPP 2.1(b)) (emphasis added).

There is also a broad exception to the obligation to maintain confidentiality and privacy, both at common law and under the privacy legislation, which justifies disclosure of serious risks to protect a third party. The federal *Privacy Act* Sch 3, NPP 2.1(ea) permits the use or disclosure of genetic information about an individual to a genetic relative in circumstances where the genetic information may reveal a serious, but not necessarily imminent, threat to a genetic relative's life, health or safety.[18]

Personal information may also be disclosed without consent in other circumstances, provided the disclosure is in accordance with guidelines approved

14 Disclosure to participants themselves is not a breach of confidentiality because breaching confidentiality involves telling people other than the person concerned. Note that the privacy legislation does not apply to patients and their relatives so those people are not restricted in passing information on to family members, provided this is for the purposes of their 'personal, family or household affairs': *Privacy Act 1988* (Cth) s.16E; *Health Records Act 2001* (Vic) s.13.

15 Australian privacy legislation includes the *Privacy Act 1988* (Cth) (which covers private organizations and individuals as well as the public sector agencies to which it initially applied); *Health Records (Privacy and Access) Act 1997* (ACT) (public and private sector); *Privacy and Personal Information Protection Act 1998* (NSW) (public sector); *Health Records and Information Privacy Act 2002* (NSW) (public and private sectors); *Information Privacy Act 2000* (Vic) (public sector); *Health Records Act 2001* (Vic) (public and private sector).

16 Under privacy legislation: see note 15 supra.

17 E.g. Health Professions Registration Act 2005 (Vic).

18 In some jurisdictions, the risk must be 'imminent' as well as serious: see *Health Records Act 2001* (Vic) HPP2; Otlowski 2007.

by the Privacy Commissioner under section 95A (Sch 3, NPP 2.1 (d)(ii)).[19] Guideline D5 of those guidelines states that a Human Research Ethics Committee (HREC) may approve research that would otherwise breach the Privacy Act if it is persuaded that 'the public interest in the proposed activity substantially outweighs ... the public interest in privacy'. Guideline D5 then lists matters for HRECs to consider in making this balance. If an organization discloses information under these guidelines without a person's consent, it must reasonably believe that the recipient of the information will not disclose it, or personal information derived from it to other people: Sch 3, NPP 2.1 (d)(iii). This provides some protection of privacy if a person's information, or the tissue from which the information has been derived, passes into other hands.

However, the biobank would be wise to take formal measures to protect itself from later liability if the information or tissue is wrongfully disclosed by a third party. The participant would not have a direct contractual relationship with the third party and, if the third party misused tissue or information derived from the participant, the participant might either bring an action against the third party for equitable breach of confidence or claim against the biobank for breach of contract. The equitable action would allege that the information was obviously confidential and should not have been revealed without consent. It would not require a contractual relationship to be proved and the biobank would not need to be involved. However, if a claim was made in contract, the biobank would be the first party sued and it would have to pursue any rights it might have against the third party in respect of the wrongful disclosure. It would therefore be wise for the biobank to include in its contracts with third parties similar provisions to those in its contracts with the participants.

Contacting Participants and Others: When, Who and How?

It would seem from the foregoing discussion, particularly in relation to statutory biobanks and implied contractual terms, that biobanks may have a legal duty, if 'research reveals information that bears on the wellbeing' of participants, to 'consider how to make that information available' to them, even if there is no duty to contact them directly. But what does this mean in practice? In particular, how severe and certain does the risk have to be to justify the effort in retrieving the information from the biobank and its records; and then 'mak[ing] that information available'? Must it involve a serious, life-threatening disorder? Must the findings be certain and not speculative? Does it require that interventions are possible? Can

19 Guidelines under sections 95 and 95A of the *Privacy Act 1988*, December 2001 at: http://www.nhmrc.gov.au/publications/synopses/_files/e26.pdf; http://www.nhmrc.gov.au/publications/synopses/_files/e43.pdf [both accessed 9 October 2009]; these guidelines apply to information held by public sector and private sector agencies respectively.

the biobank take account of the number of people likely to be affected and the time and cost required in contacting people?

The second issue is who should be informed. As noted earlier, Australian courts have not recognized a general duty to warn people who have a risk that is known to a health professional but not to them, even if there is a risk of serious harm. However, the NHMRC *National Statement* (NHMRC 2007) and the federal privacy guidelines discussed earlier envisage that it may be necessary to 'consider how to make that information available'. This obligation seems not to be limited to participants and it may possibly extend to close blood relatives. If that is the case, the participants should first be counselled to contact the relatives and told that if they do not do so, the participant's personal information may be disclosed to relatives without the participant's consent. This will be a decision to be made in each case but it would seem generally to be advisable to warn the person so as to give advance notice of matters that will be discussed in the family. If the circumstances fall within the legislation[20] (i.e. a serious risk; genetic relative), the person's refusal to consent to the information being given to relatives may be legally overridden. Similarly, if the biobank has given an undertaking when the material or information was first collected that it would not disclose any information without consent, that undertaking could possibly also be legally overridden, even if it constituted a contractual undertaking that induced the person to agree to participate in the biobank. That issue has not arisen in practice in Australia but the policy in the recent amendments to the Privacy Act 1988 (Cth) Sch 3, NPP 2.1(ea) might be argued to override a contractual obligation to maintain confidentiality where there is a serious risk to relatives that could be averted or minimized by a warning; or requiring confidentiality to be maintained in such circumstances might be regarded as contrary to public policy.

Having decided who should be informed, the next issue for the biobank would be how to make contact. The biobank could report in general terms on its website, recommending that anyone who may be affected by a particular finding should contact their own doctor for advice and a test if necessary. This option has the attraction of avoiding a breach of confidentiality and being relatively simple to implement, but one may question whether the people who need to know the relevant information will be regularly consulting the web site for updates. Also, there may be difficulties in phrasing information on the web site in a way that will not cause undue alarm among people who are not at risk, especially if they are reading the information when they do not have anyone with them to answer questions and provide support.

An alternative form of communication would be for the biobank to inform doctors directly (as in the Joondalup Family Health study in Western Australia)[21]

20 For example, *Privacy Act 1988* (Cth) Sch 3, NPP 2.1(ea).

21 Participants in this study may 'choose to receive notification of any test results that indicate a potentially significant health issue … to participants and/or their GPs', but not 'genetic feedback': see Project website, Ethical FAQs, question 8: http://www.jfhs.org.au/faq-ethical.html#faqE2 [accessed 9 October 2009].

by placing a note within the patient's electronic medical record. This may be facilitated as electronic health records become more widely used. This may provide a relatively simple method of notifying findings directly to medical practitioners caring for relevant patients and even their relatives. In the Joondalup Family Health study participants would know that this type of feedback would occur as part of their participation in this research study. Therefore, it would not be a complete shock if they were asked by their doctor to attend for an additional consultation. However, there is still considerable debate as to whether feedback should be channelled through medical professionals or if the participant should be contacted directly, which is dependent upon the type of information that will be returned.

The most specific response to participants – and one mentioned specifically in the *National Statement* (NHMRC 2007), is 'individual notification' – a personal letter to people affected, that is, those recruited with a disease, rather than 'healthy volunteers' (though as suggested earlier, the legal duties arising from a clinical relationship may be different from those arising in other research). The letter could be in fairly general terms with an invitation to seek an appointment for more detailed discussion. The ethics committee overseeing the biobank could be consulted about methods of contacting participants.

Legal advisers and insurers

If there is a significant finding, biobanks may want to contact their legal advisers and insurers. Indeed, prompt notification of information that may be relevant to potential claims is likely to be a term of the insurance contract. An insurer may be concerned about the cost of potential claims if participants are not notified of a significant finding, but such claims are fairly unlikely, especially if participants are informed soon after the finding is made. The biobank will be concerned about possible reputational cost if it does not reveal the finding, which is less likely to be covered by insurance than personal claims by participants. However, the reputational cost is likely to be greater if it should ever become known that the biobank had important medical information that could have prevented a serious risk eventuating and it did nothing, than if it is known to have breached confidentiality by warning participants and close blood relatives.

Conclusion

There appear to be no direct provisions in Australian law that require the custodians of biobanks and other large scale repositories of human tissue and medical information to warn participants when a 'significant finding' is made in research from those sources. This applies equally to warning other people, such as close blood relatives of the participants, who may also be at risk. However, if the biobank chooses to provide feedback to participants and relatives, it will probably not be liable in civil litigation for breach of confidentiality, or complaints under privacy or disciplinary legislation.

If the biobank, or third parties who later acquire tissue or information from the biobank, make significant findings and do not disclose them to participants (and possibly their blood relatives and others who may have a similar risk), it is possible that those people may claim in contract or tort (property arguments are unlikely to be considered). An action in contract might be successful if there was a term requiring the biobank to provide feedback to participants. However, in the absence of such a term, it is likely that the implied obligation of the biobank under the contract would be to take reasonable care and, by reference to the *National Statement* (NHMRC 2007), that would be interpreted to mean that it would have a duty to *consider* how to make information available, rather than to directly warn participants (and others). The same appears to be true with biobanks established by statute.

Claims by participants in tort for failure to warn are unlikely to be successful. Australian law has not recognized a general duty to warn and even if a breach of the biobank's duty to take reasonable care in providing feedback could be established, it would be difficult for the participants to prove that their injury or loss was caused by that breach and not by their genetic predisposition.

Acknowledgements

- Australian Research Council (ARC) – funding.
- Centre for Law and Genetics, University of Tasmania.

Bibliography

Australian Law Reform Commission. 2008. *For Your Information: Australian Privacy Law and Practice* (ALRC 108).

Cho, M.K. 2008. Understanding Incidental Findings in the Context of Genetics and Genomics. *The Journal of Law, Medicine & Ethics*, 36, 280-285.

Johnston, C., Kaye J. 2004. Does the UK Biobank have a Legal Obligation to Feedback Individual Findings to Participants? *Medical Law Review*, 12(3), 239-267.

Laurie, G. 2002. *Genetic Privacy, A Challenge to Medico-Legal Norms*. Cambridge: Cambridge University Press.

Morgan, D. 2001. Where Do I Own My Body (and Why?), in *Issues in Medical Law and Ethics*. London: Cavendish, 83-104.

National Health and Medical Research Council. 1999. *National Statement on Ethical Conduct in Human Research*.

Otlowski, M.F.A. 2007. Disclosure of Genetic Information to At-risk Relatives: Recent Amendments to the *Privacy Act 1988* (Cwlth). *Medical Journal of Australia*, 187, 398-399.

Skene, L. 2002a. Proprietary Rights in Human Bodies, Body Parts and Tissue: Regulatory Contexts and Proposals for New Laws. *Legal Studies*, 22 (1), 102-127.

Skene, L. 2002b. Arguments against People 'Owning' their Own Bodies, Body Parts and Tissue. *Macquarie Law Journal*, 2, 165-176.

Van Ness, B. 2008. Genomic Research and Incidental Findings. *The Journal of Law, Medicine and Ethics*, 36(2), 292-297.

Wolf, S.M., et al. 2008. Managing Incidental Findings in Human Subjects Research: Analysis and Recommendations. *The Journal of Law, Medicine and Ethics*, 36, 219-248.

Wolf, S.M., Paradise, J. and Caga-anan, C. 2008. The Law of Incidental Findings in Human Subjects Research: Establishing Researchers' Duties. *The Journal of Law, Medicine and Ethics*, 36(2), 361-383.

Zimmerman, S.V. 2005. Translating Ethics into Law: Duties of Care in Health Research Involving Humans. *Health Law Review*, 13(2/3), 13-18.

Chapter 11
Ensuring Participant Privacy in Networked Biobanks

Atieh Zarabzadeh, R. William G. Watson,
Geoff Bradley and Jane Grimson

Biobanks, biorepositories or bioresources, terms used interchangeably in the literature, denote resources for storing biological samples and data to support the discovery of biomarkers, therapeutic targets and the underlying causes of diseases (Betsou et al. 2004). The importance of these repositories is that they store and maintain quality controlled and standardized samples (Patel et al. 2005), which can be used by many different researchers for many different purposes. The Irish Prostate Cancer Research Consortium (PCRC) has developed a research biobank for prostate cancer tissue, blood, urine and DNA. Ensuring the confidentiality of participants' data is one of the major requirements of biobanks and their associated data management systems. The PCRC biobank has adopted best practice based on the requirements of national legislation and regulations regarding data protection. This chapter will provide an overview of participant confidentiality and the methods used to protect data which identifies participants. It will discuss the challenges in applying the principles of Data Protection legislation and discuss how these are addressed by the PCRC biobank and the development of its associated Biobank Information Management System (BIMS). The infrastructure of the PCRC BIMS and its sample management system will be described. Finally, an explanation of how the PCRC biobank adheres to best practice is provided in the conclusion.

The Irish Prostate Cancer Research Consortium (PCRC) Biobank

The Irish Prostate Cancer Research Consortium (PCRC) has developed a research biobank focusing on prostate cancer. The PCRC biobank is a federated multi-location multi-purpose biobank consisting of four major Irish hospitals as collection sites and two research institutes with responsibility of carrying out studies for detection of biomarkers. It is planned to add more collection sites and research institutes to the consortium in the future. The Irish PCRC biobank facilitates collections of tissue, blood and urine samples by clinical and nursing staff as well as performing omic techniques by scientists and researchers. Samples and data are collected from participants every six months and are linked longitudinally over time to allow analysis of regression or progression of the disease. Details of the infrastructure of the PCRC biobank will be discussed later in this section.

The aim of biobanks is to create a resource that can be used to understand the underlying molecular causes of diseases. This will in turn lead to more accurate diagnosis, better targeted treatments with fewer side effects, and even disease prevention strategies. In order to do this, a biobank must store and maintain quality controlled and standardized samples (Patel et al. 2005), but also allow researchers to access them in a way that protects patient privacy. The International Evaluation of Swedish Bio-repositories (Sorensen et al. 2005), defines a biobank as:

> [a] long-term depository of biological samples from an identifiable human population. The content of the depository must be of such quantity and quality that it is suitable for later biomedical analysis in epidemiological and clinical research for individual clinical purposes.

Biobanks facilitate continuous collection of data and information over extended periods of time in order to maximize the potential for knowledge discovery and to reduce the costs of future research (Holland et al. 2005). They require identifiable samples and information. Ölund, Lindqvist and Litton (2007) have even suggested that the importance of information about the participants themselves is of equal or even higher value than the information contained in the samples.

The Legal Requirements

The British Medical Association (BMA) has published guidelines for ethics in healthcare. These guidelines state that research should be approved by 'an appropriate research ethics committee in compliance with clear confidentiality rules' and that good research practice requires that data is either anonymized or that coded identifiers are used for personal data (Nathanson et al. 1999). Participant consent should be sought only after participants have been informed of the precise purpose of data collection and of their rights. They may refuse to give their consent (Nathanson et al. 1999). Neither the UK (*Data Protection Act 1998* (UK)) nor the Irish Data Protection Acts (Data Protection Commissioner Ireland 2007) include any specific rules and regulations for biobanks. However the eight principles regarding data safety included in both Acts apply equally to biobanks as to any other repositories. In brief, the eight principles are that personal data should be:

1. fairly and lawfully processed;
2. processed for limited purposes;
3. adequate, relevant and not excessive;
4. accurate;
5. not kept for longer than is necessary;
6. processed in line with data subjects' rights;

7. secure; and
8. not transferred to countries without adequate protection.

These principles limit the scope of data availability in order to enhance integrity, minimise unauthorized access, and prevent data disclosure to inappropriate individuals. They also ensure that data is only used for the specific purpose for which it was collected. In addition to these eight principles, the Data Protection Commissioner in Ireland has published guidelines for research in the health sector (Hawkes 2007). According to these guidelines if pseudonymized data, e.g. replacement of identifiable data with initial or coded data, are not sufficient for research purposes, consent should be obtained from the patient.

The Challenge of Protecting Participant Privacy in the PCRC Biobank

Ensuring the confidentiality of participant data at all times is an essential aspect of biobank operation. Firstly, participants must be fully informed of the purpose of the study, and of how their data and samples will be used and contribute to current and possibly future studies. Before any sample is obtained from a participant in the Irish PCRC biobank an information leaflet is provided to them by the dedicated research nurse who is also available to address any questions. The details of the study explaining the purpose, the information that will be collected, why this particular person has been chosen, and who is organizing the study are provided in the information leaflet. Participants are assured of their anonymity and are given the right of withdrawal at any stage of the research. A consent form is signed once they agree to participate. In case of withdrawal, all of their specimens are located and destroyed. In the event that a participant becomes incapable of making decisions or dies, a close relative is in charge of making decisions regarding on-going consent.

For the successful development of biobanks, it is vital that they can guarantee the confidentiality of participants participating in the study and that there will be no unauthorized disclosure of personal data. However, many studies require the ability to collect samples from individual participants over time and these therefore need to be linked together. Furthermore, the scientific value of a specimen or a data set is mostly determined by the clinical, short and long-term outcome data that supports them (Isabelle et al. 2006). Information such as age of the participant, clinical history and long-term outcome require identifiable attributes. Access to participant information that directly identifies an individual participant such as name, age and medical record number must be restricted to those who have direct contact with the participants at sample collection time. The data must then be subjected to a process of de-identification which removes personal data, replacing, for example, date of birth with age (year only) at the time of collection, or medical record number with a study number.

It is particularly important to protect the privacy of individuals when samples are being transferred between institutions. In the PCRC biobank the samples are collected from identified participants in each of the four hospitals and then sent to one of the two research institutes for processing. The researchers in the research institutes need to access the clinical and sample data that have been collected at the hospitals, without being able to access any identifiable data.

The National Bioethics Advisory Commission (NBAC www.bioethics.gov) has described four methods for concealing confidential information for use in biobanks in a guideline published in 1999 (National Bioethics Advisory Commission 1999). The American Society of Human Genetics (www.ashg.org) has defined similar methods using slightly different terminology (Godard et al. 2003). The four approaches are as follows:

1. Unidentified or anonymous samples are samples for which personal identifiable information has not been collected or if collected, is not maintained. Therefore they cannot reveal any confidential data.
2. Unlinked or anonymized samples are stored without identifiers or codes that would link samples to other identified or identifiable specimens or individuals. These samples are supplied to researchers with no identifiers or confidential information. The biobank or 'a disinterested party' maintains individuals' identifiable information and hence linking this category of samples to identifiable information should not be possible.
3. Coded or linked or identifiable or de-identified samples do not disclose any individual's identifiable information; however they are coded by the biobank or its agent before being made available to external scientists and researchers. The code can be used only by the coding body in order to associate individuals' identifiable data with the specimens. This approach is referred to as pseudonymization.
4. Identified samples are those that are made available to researchers which are accompanied by the individuals' identifiable information such that linking samples with the individuals from whom the specimens are collected is straight forward.

The process of de-identification is the most suitable approach for maintaining participant privacy. In contrast to fully anonymized or anonymous data, de-identification is reversible by methods of re-identification.

The Karolinska Institutet (KI) biobank, a well facilitated biobank in Sweden, has developed its BIMS by integrating sample, phenotype, genotype and medical record data in a database. The integration of these data facilitates complex searches to be carried out on the biobank samples supporting knowledge discovery (Ölund et al. 2007). The query searches are possible via the biobank website and its portal from ki.se/kibiobank. These four categories of data are stored in the database of the BIMS after collection, de-identification and preparation. Then the integrated information is made available to the researchers. The BIMS system that IBM has developed for the KI biobank only allows data and samples to be accessed in accordance with the participants' wishes as set out in the consent form (Schreier 2008).

The Cooperative Prostate Cancer Tissue Resource (CPCTR) in the US is a distributed biobank comprising of four sites. CPCTR has employed a method of de-identification of data at local sites prior to transfer of the data to the central database. Re-identification is done through a 10-digit randomly generated number that maps the de-identified data of central database to the identifiable data of the local databases (Patel et al. 2005). The mapping table is maintained by an honest broker, the tissue bank trustee.

The Irish PCRC where personally identifiable data is stored only at the collection sites uses the same method of data management and de-identification as the CPCTR. The local participant identifier is replaced by a globally unique study number prior to transmission to the central repository. The research nurse at each of the hospital sites fulfils the role of the honest broker and maintains the mapping between local and global identifiers, thus enabling linkage of individual participant records over time. The research nurse requires identifiable information as they are in direct contact with the participants at the time of consent, sample collection and subsequent follow-ups.

The PCRC Biobank Information Management System

A critical initial step in developing a biobank is to find a way to store both molecular (genotypic) and phenotypic information, and then to build a relationship between them and the corresponding samples through a database. A BIMS is responsible for coordinating the communication and information management for biobanks. The BIMS makes the link between the participant and sample data and the actual biological samples that are stored in designated freezers. A BIMS is defined as:

> … a middleware system that will handle communication between several other systems. … The system will be responsible for a range of tasks, including: Middleware functionality. … Result storage. … Participant consent handling. … Sample and information management. … (Litton 2004)

The BIMS must be able to handle constantly updated data, control access to data with a user-friendly interface, and finally allow query searches on the data. Most of the effort of designing a BIMS focuses on defining, structuring and standardizing the information collected, such that they are comparable and of standard quality (Ölund et al. 2007). Maintaining participant confidentiality is a critical function of the BIMS. The PCRC biobank has deployed a BIMS that was developed using the Distiller software from Slidepath (www.slidepath.com/) and is used to support the biobank across the hospitals sites and sharing of data across the research institutes.

The Irish PCRC biobank consists of four major components: the four participating hospitals with their Clinical Resource Centres (CRC), the two research institutes, a sample tracking and identification system, and the central database. Data and samples are collected from the participants in hospitals and initial processing, including quality assurance, are carried out in their corresponding CRC. After the initial processing phase, de-identified samples are stored in the biobank's designated freezers and are made available to the research institutes. The central database stores de-identified data from each unit. This includes participants' clinical information, biopsy, radical prostatectomy, biochemistry, participant outcome and information on collection of three main types of biological samples: tissue, urine and blood. In accordance with best practice, no identifiable information about participants is stored on the central database. Only de-identified data is exported from the local databases at the hospital sites to the central database. Figure 11.1 shows the units involved in the PCRC BIMS.

The tracking and identification system is responsible for ensuring that samples are accompanied with related data and information throughout all stages of processing. It enables the BIMS staff to locate the sample anywhere within the system without revealing any identifiable data.

Hence, the PCRC uses a distributed database system, where personal data such as date of birth of participants, name and medical record number are stored in a separate local hospital database which is linked to the central database by a study number, which is maintained and accessed locally. This allows the individual hospitals to follow up the progression of the disease among the participant population and populate the central database with emerging data. The central repository and the local databases at the collection site together comprise the biobank information management system. These local databases at the hospital sites are operated in accordance with the respective hospitals' security and confidentiality policies and are accessible only by the research nurses and clinicians at those sites who are already aware of participants' identities. The local databases export de-identified data to the central database. Figure 11.2 shows the relationship between the local and central databases. The PCRC Biobank system has been reviewed and approved by the Irish Data Protection Commissioner. In addition to the de-identified clinical and sample data exported from the local hospital databases, the central database also contains the results of omic techniques and other data from the research institutes. Access to the central database is restricted to authorized, registered users of the PCRC. Depending on their predefined privileges, they may have the right to edit the data. The PCRC maintains a complete audit trail of all accesses to the central database. These logs can be used to track back and identify any unauthorized access to the database as well as to monitor operations performed by each account holder.

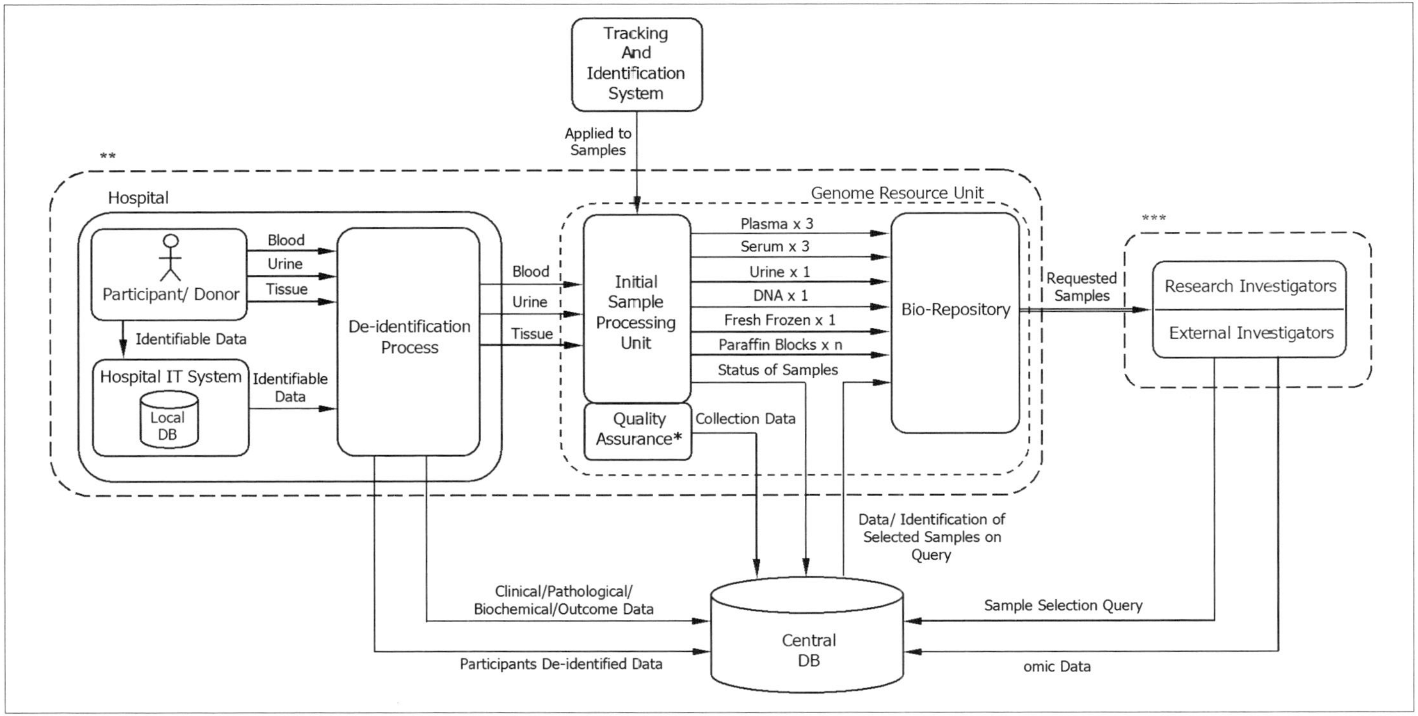

Figure 11.1 The Irish PCRC BIMS

Note: * Based on time and adherence to SOPs; ** Hospitals which consist of the Hospital itself and its GRU; *** There are two research institutes at the moment: IMM and Conway Inst.

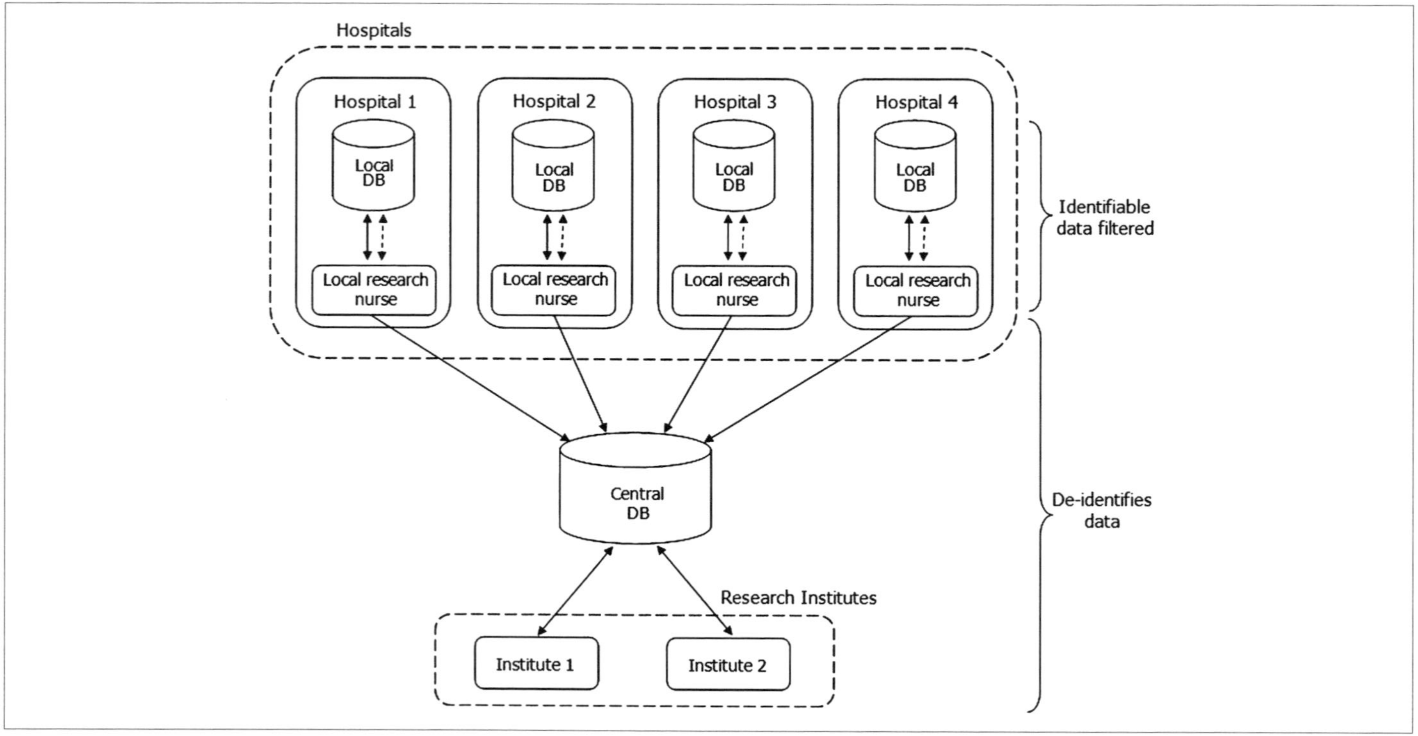

Figure 11.2 The site plan of Irish PCRC Biobank

Note: ◄– – – – – –► = Patient identifiable data; ◄————► = De-identified data.

Samples and data coding

To identify data, each collection site has introduced an Institute Study Number (ISN) for each participant. This ISN consists of a three-character abbreviation of the name of the hospital, followed by an arbitrary number unique to the participant. This identifier is only used for identifying data and participants within the domain of the corresponding collection site (hospital). The ISN reveals the collection site and hence the hospital at which the participant was treated. This is regarded as confidential data as it could assist in identifying individual participants in the event that additional information, such as date of surgery, is known. A unique PCRC Study Number is introduced to prevent recognition of participants within BIMS as a distributed system. Therefore, the PCRC Study Number is used as a global identifier of any data and sample within the Irish PCRC biobank and is the key identifier which can be mapped back to the original participants' data via the local database of hospitals.

Conclusions

The key requirements for best practice for ensuring the confidentiality of participant data and samples and which ensure adherence to legislation and ethical considerations are summarized below. The solution adopted by the PCRC to ensure conformity with best practice is explained for each of these requirements.

1. The study must have approval from the individual ethics committees of the participating hospital collection sites. The method of collecting and handling sensitive and de-identified data must adhere to national data protection regulations, and the biobank must be registered with the appropriate body.

Studies that are carried out on the PCRC biobank samples and data are approved by the individual hospital ethics committees and Data Protection Commissioner. Some of the criteria taken into consideration in the approval process are consent management, role based access to data, Standard Operating Procedures (SOPs), audit trail facilities and separation of confidential data.

2. A strategy should be in place to record and manage participant consent. No sample and data shall to be collected or stored without first obtaining participant consent.

SOPs regarding consent and sample collection have been developed. Clear and unambiguous information leaflets and consent forms are given to the participants before any sample is obtained. Participants have the right to withdraw their consent at any time. The Irish PCRC biobank does not fully anonymize data and samples so they can be removed in the event that consent is withdrawn.

3. No personally recognizable information will be revealed or go beyond the hospital system or be seen by people beyond hospital staff. Participants' identities must be considered as highly confidential data and shall not be made available to unauthorized parties.

Participants' confidential data is stored on a separate system at the collection site and is only available to research nurses who are already aware of participants' identities and are bound by the confidentiality agreements made within the hospital work place.

4. Standard Operating Procedures (SOPs) are required to be in place for every task being carried out in the biobank to make sure no additional data is gathered beyond that which is required for the study or consented by the participant. SOPs for recruiting participants, consenting participants and sample collection, processing, storage and retrieval are required to be developed and adhered to. Samples and data quality controls are met through SOPs. Sample processing duration and factors that influence their quality are recorded according to SOPs.

Dedicated SOPs for all procedures from consent management to sample collection, storage, retrieval and tracking are developed and adhered to.

5. Data should be stored in databases that provide appropriate levels of security. Electronic records are safe in terms of preventing loss and unauthorized access, easy to query and search, and are much simpler to manage than paper records. Personal data is stored on machines with limited access in terms of physical security, i.e. locked offices with restricted access, and in terms of access control, through the use of a login/password. These machines are either completely disconnected from a network, or are connected to the secure hospital network. In either case, clear procedures are required to ensure these measures are followed.

Central and local databases have been developed and their content and management approved by the relevant bodies. The local databases, which contain personal information, are on very limited access and only to biobank staff who are already aware of participants' identities. All users of the central database have access passwords and usage is monitored and recorded. Also the central database is designed to minimize accidental upload of data, e.g. use of open text fields, into which users can enter any data.

6. Authorization and authentication of system users need to be carefully managed. Privileges must be defined precisely and access should be permitted on a strictly need-to-know basis. There are guidelines available for defining safe user accounts, passwords and access control (role based access).

Roles and privileges for each of the users are defined for local and external databases in the PCRC biobank.

7. Operations performed on participants' records should be monitored and logged for possible future needs. An audit trail of all accesses should be maintained.

Logs of central database operations are maintained.

8. It should be possible to trace data and samples back to the individual participant but this function should be restricted only to authorized parties under strictly controlled circumstances.

Tracking data back to the participant is only possible via the local database which in turn is available only to authorized personnel.

9. It is important that biobanks carry out regular reviews/audits of security and data protection procedures to ensure that best practice is being maintained of all times. Also the best practice will need to be updated as new technologies or potential security threats emerge.

Regular security audits are carried out on the PCRC BIMS and its central database.

The implementation of the BIMS according to the procedures described in this chapter results in a secure and fully confidential system which could potentially be made available across a shared network of co-operating biobanks in the future. This would only be possible when a globally standardized method of data collection and storage is made available with every biobank following identical SOPs. Only in this way would it be possible to ensure that data are of the same quality.

Developing a federated network of biobanks represents the first step towards worldwide sample collection and linking BIMS from different biobanks is the next challenge if we are to fully harness the value of such biobanks (Ölund et al. 2007). A network of biobanks will allow exchange of large sample and data sets from a wide variety of biological and geographical backgrounds and hence more complete collections of samples and data will be possible. This offers the potential to accelerate knowledge discovery significantly by greatly increasing the size of the study population for both discovery and most importantly for validation of novel biomarkers. The model presented here could be used as a basis for the development of biobank networks in the future.

Acknowledgments

The Prostate Cancer Research Consortium is funded by Cancer Research Ireland and Health Research Board, IT infrastructural grant (HRB). A.Z. is funded by

Irish Research Council for Science, Engineering and Technology and Hewlett-Packard.

Bibliography

Betsou, F. et al. 2004. *BIOBANKS: Accelerating Molecular Medicine, Challenges Facing the Global Biobanking Community*. [Online] Available at: http://www-03.ibm.com/industries/global/files/Biobanks_Accelerating_Molecular_Medicine.pdf?re=healthcare&sa_message=title=biobanks:_accelerating_molecular_medicine [accessed 24/02/09].

Data Protection Act UK. 1998. *Data Protection Act*. [Online] Available at: http://www.opsi.gov.uk/Acts/Acts1998/ukpga_19980029_en_1 [accessed 23/02/2009].

Data Protection Commissioner Ireland. 2007. *The Data Protection Rules*. [Online] Available at: http://www.dataprotection.ie/ViewDoc.asp?DocId=21 [accessed 26/11/2007].

Godard, B. et al. 2003. Data Storage and DNA Banking for Biomedical Research: Informed Consent, Confidentiality, Quality Issues, Ownership, Return of Benefits. A Professional Perspective. *Eur J Hum Genet*, 11(S2), S88-S122.

Hawkes, B. 2007. *Data Protection Guidelines on Research in the Health Sector*. [Online] Available at: http://www.dataprotection.ie/documents/guidance/Health_research.pdf [accessed 24/02/2009].

Holland, N.T. et al. 2005. Molecular Epidemiology Biomarkers – Sample Collection and Processing Considerations. *Toxicology and Applied Pharmacology*, 206(2), 261-268.

Isabelle, M. et al. 2006. TuBaFrost 5: Multifunctional Central Database Application for a European Tumor Bank. *European Journal of Cancer*, 42(18), 3103-3109.

Litton, J.-E. 2004. *A Biobank Information Management System: The Importance of Data Integration*. [Online] Available at: http://www-03.ibm.com/industries/global/files/Biobanks_Accelerating_Molecular_Medicine.pdf?re=healthcare&sa_message=title=biobanks:_accelerating_molecular_medicine [accessed 24/02/09].

Nathanson, V. et al. 1999. *Confidentiality and Disclosure of Health Information, Guidance from the BMA's Medical Ethics Department*. [Online] Available at: http://www.eastmidlandsdeanery.nhs.uk/document_store/12142148621_confidentiality.pdf [accessed 24/02/2009].

National Bioethics Advisory Commission. 1999. *Research Involving Human Biological Materials: Ethical Issues and Policy Guidelines*. N.B.A. Commission: 138.

Ölund, G. et al. 2007. BIMS: An Information Management System for Biobanking in the 21st Century. *IBM Systems Journal*, 46(1), 171-182.

Patel, A. et al. 2005. The Development of Common Data Elements for a Multi-institute Prostate Cancer Tissue Bank: The Cooperative Prostate Cancer Tissue Resource (CPCTR) Experience. *BMC Cancer*, 5:108.

Schreier, P. 2008. *Biosample Storage*. [Online] Available at: http://www.scientific-computing.com/features/feature.php?feature_id=195 [accessed 11/12/2008].

Sorensen, T.I.A. et al. 2005. *The International Evaluation of Swedish Biobanks*. [Online] Available at: http://www.biobanks.se/documents/Evaluation%20of%20Swedish%20biobanks%20Final.pdf [accessed 24/02/2009].

Chapter 12

hSERN: A Tool to Help Researchers with the Legal Requirements of Cross-border Exchange of Biological Material

Emmanuelle Rial-Sebbag, Aurélie Mahalatchimy, Dennis Chartier and Anne Cambon-Thomsen

One of the consequences of the advances in biology, biotechnology and biomedicine is a growing need to exchange human biological materials (HBM) between biobanks and research collaborators. While biobanks are designed to make HBM more accessible to a large number of researchers, they do not necessarily address the procedural issues of how to transfer HBM between biobanks located in different countries. Our experience is that the regulations that apply to the importation or export of HBM is a key question for researchers when collaborating (Cambon-Thomsen, Rial-Sebbag and Knoppers 2007) and if not executed properly this can be a major stumbling block to research progress. The tool we have developed within the Global Allergy and Asthma European Project, Ga^{2}len (http://www.ga2len.net/), is focused on the requirements provided by legislation for the import and export of biological samples of human origin, which are partly embedded in measures for the protection of public health. Although these rules must be respected by the researchers and/or their institutions, researchers also use different kinds of instruments to organize their international collaborations, such as Material Transfer Agreements (MTA), which may contain requirements in addition to the legally required ones. This tool only addresses the issue of shipping human samples for research use – the question of shipping human samples for therapeutic use is not treated. In this chapter and in the corresponding tool, we restrict our focus to legal requirements and do not address the question of MTAs as these can be specific to each partner or institution. The purpose of this chapter is to describe the Human Sample Exchange Regulation Navigator (hSERN, www.hsern.eu/) in order that others can learn from our experience.

The Problem of Cross-border Transfers

Increasingly, we are seeing the encouragement of collaborations between researchers in different countries of Europe but also at a global level, in order to further scientific progress. This is being done by funding bodies that reward research projects that bring in researchers from different member states within Europe. This is also mirrored in new regulations or professional recommendations that encourage open access to databases and the sharing of samples (Kaye 2009). For example, the International Declaration on Human Genetic Data adopted in 2003 (UNESCO 2003) stressed this international cooperation in Article 18 which says:

> Article 18: Circulation and international cooperation
>
> (a) States should regulate, in accordance with their domestic law and international agreements, the cross-border flow of human genetic data, human proteomic data and biological samples so as to foster international medical and scientific cooperation and ensure fair access to this data. Such a system should seek to ensure that the receiving party provides adequate protection in accordance with the principles set out in this Declaration.
>
> (b) States should make every effort, with due and appropriate regard for the principles set out in this Declaration, to continue fostering the international dissemination of scientific knowledge concerning human genetic data and human proteomic data and, in that regard, to foster scientific and cultural cooperation, particularly between industrialized and developing countries.
>
> (c) Researchers should endeavour to establish cooperative relationships, based on mutual respect with regard to scientific and ethical matters and, subject to the provisions of Article 14, should encourage the free circulation of human genetic data and human proteomic data in order to foster the sharing of scientific knowledge, provided that the principles set out in this Declaration are observed by the parties concerned. To this end, they should also endeavour to publish in due course the results of their research.

For scientists, the transfer of data and samples is seen as adding value to collaborations and is essential for the scientific work. This key point has been clearly addressed by the European Society of Human Genetics in its recommendations on data storage and DNA banking for biomedical research, technical, social and ethical issues adopted in 2001. The Recommendation relating to Access/Sharing and Ownership stresses that 'while protecting confidentiality the free circulation and availability of genetic information and samples for research should be promoted' (European Society of Human Genetics 2003).

Research collaborations where HBM are circulated and exchanged within one jurisdiction raise different legal issues to transfers that occur between collaborators in different countries (Zika et al. 2008). In the case where transfers are carried out in one country, the legal issues are focused on institutional agreements and the legal requirements for transfer that are determined at a national level. In the case of international research projects, such as projects funded by the European

R&D framework programmes, different laws that apply in each jurisdiction are involved. Researchers normally exchange results, samples and data according to a consortium agreement. A researcher, for example, could need specific HBM which are only available in another country. Thus, to import the specific needed HBM, researchers will have to respect the law of their own country that applies to the import of HBM, and their colleagues in the exporting country will also have to follow the legal requirements that apply to the export of HBM. Moreover, international law and European law also have requirements for the exchange of HBM, which must be respected.

In view of this multiplicity of enforceable laws and guidelines, it is difficult for researchers and lawyers to know what should be done and what principles should be followed. There is a concern that these legal requirements may be an obstacle to the development of sharing of HBM, and could impede the progress of science. In considering whether to exchange HBM for any kind of research, two fundamental principles must be balanced: the respect of the human dignity and the freedom of research. These are well known in the field of bioethics and embodied in the law. Currently, it is evident that some researchers' knowledge concerning legal provisions is limited and that many ignore the regulations and the sanctions that could apply to any infringements of the law. It was to address this need, that the hSERN tool was developed.

What Led to a Web-based Tool?

In 2006, the 'Genetics and Society' Platform from the 'Genotoul' (http://www.genotoul.fr/index.php?id=74) was asked by a European Excellence Network called GA²LEN to work on the legal aspects for exchanges of biological samples within their research project. GA²LEN (Global Allergy and Asthma European Network) is a consortium of leading European research centres addressing the growing public health concern of allergic diseases. This project was funded by the European Union under the 6th framework programme and commenced on February 1, 2004, with a duration of 60 months. The work package 2.9, 'Genetics, genomics and post-genomics', led by Francine Kauffmann from the 'French National Institute of Medical Research and Public Health' (INSERM), focuses on methodological and organizational aspects which promote collaborations between researchers. In this project, a general lack of knowledge regarding the legal aspects of the exchange of biological for research was noted as well as the heterogeneity of the sources of such information and the fact that they were widely dispersed. Very few partners were using material transfer agreement forms when shipping biological samples to another laboratory or country.

The Genetics and Society Platform is a societal platform directed by Anne Cambon-Thomsen, in the framework of a Genopole, a group of collaborating centres in Toulouse (Genotoul). The Genetics and Society Platform aims to provide information to researchers on bioethical questions that arise in the context of genetic research conducted in the health sector. This is a multi-disciplinary advisory service that assists researchers with the legal and ethical questions linked to bioethical aspects of genetic research, as well as the wider social implications of the research.

In order to satisfy GA²LEN's request, the Genetics and Society Platform decided to make available in a practical way the principles and general legal and ethical texts related to the conservation, management and exchange of samples with texts of different binding levels. Rather than producing a written report, it was decided to conceive and implement a web based tool, called hSERN, which facilitates the addressing of a numbers of requests, with respect to different countries, on the issue of the regulatory aspects of exchanging HBM across borders. This tool permits those registered for the service to get information on the conceptual as well as practical legal aspects, of cross-border exchanges of HBM for research purposes.

Our Methodology

Our first task was to identify the frequent questions faced by researchers, regarding the practical aspects of importing and exporting HBM. The main questions that we identified were: What are the documents and forms that have to be filled in? Which authority do they have to be returned to? What are the issues around consent that emerge from the use of HBM?

Second, we identified the relevant legal texts in the each of the countries involved in the GA²LEN project. For the time being and largely for practical reasons, we have restricted this tool to the 16 European countries involved in GA²LEN (although if we included all of the collaborating centres in GA²LEN this would boost this number to 25 countries). A proof of concept was done with two countries, France and the UK. The study will subsequently be extended to two other countries, Belgium and Spain. The intention is to then further expand the study after intensive practical validation by researchers in these four countries. Research on and analysis of the law was undertaken by jurists from the Genetics and Society Platform. We also sought legal experts in each of the selected countries to validate the information and our analysis.

In order to do this, we also had to assemble a team of people with the requisite skills. Lawyers and biomedicine scientists were essential, not only to search documents but above all to understand the issues at stake from the researcher's point of view and to incorporate them in the tool. This work was the first step in designing the database, which is the major instrument of the tool. In addition, a strong collaboration with computer specialists was essential to construct the questions tree, to program and design the website, to conceive and to facilitate the easy updating of the database and to organize the queries. The figures presented here are reproduced from the prototype of the tool; we are still working on improving the design as the tool is now available to the scientific community.

How does hSERN Work?

The queries

Before being granted access to the tool, the researcher must complete registration and log in information, using a form. The purpose of this is to develop statistics on the uses of the tool, with a view to adjusting its content. The original name chosen for the prototype was SERN, but this had to be changed to hSERN in order to more accurately describe the content, which only addresses human biological samples, but also to fit with the available web domain names.

After log in, the country of export and the country of import of the HBM are selected. The software will then select the respective requirements for each specified jurisdiction. We chose a common case study, that of sending HBM from France to the UK, in order to test the process. The relevant rules were thus the French rules on exportation and the English rules on importation. Having selected the countries in question, the database then displays the legal requirements in each jurisdiction for the researcher.

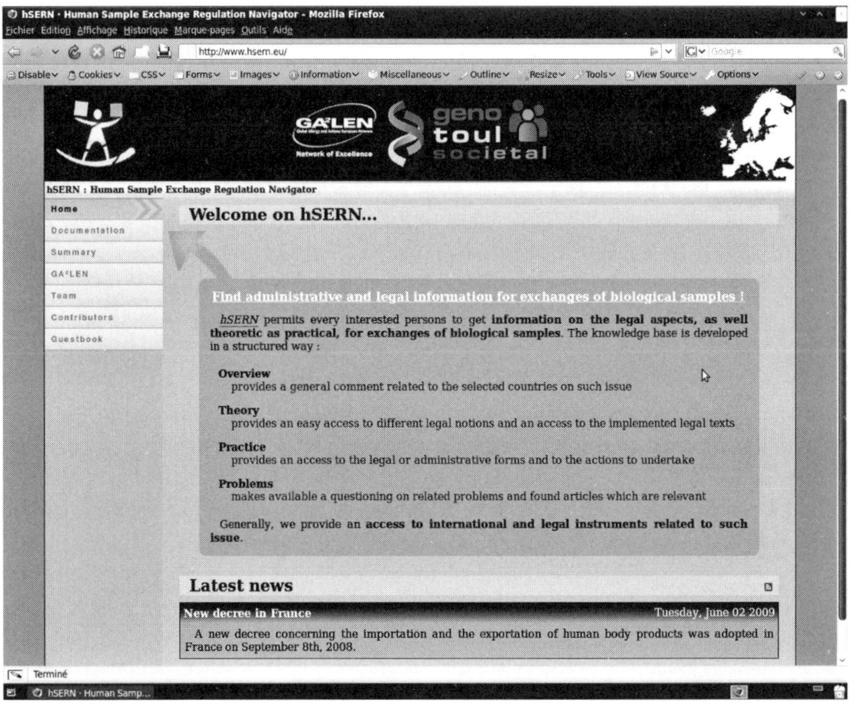

Figure 12.1 hSERN home page

The information provided by the tool

Four categories of information are available within hSERN: Overview, Theory, Practice and Issues. As shown in Figure 12.2 (also from the prototype), our test case study was the export of HBM between France and the UK. The respective sections of hSERN provide the following information:

- 'Overview' provides a general comment related to the selected countries on the exchange of HBM. This box specifies the nature of the instruments to be followed (law, regulation, guidelines). For example, '*France has a specific framework to organize import/export of HBM for research. It consists of law and other regulations. They indicate the French authority delivering an authorization and the different requirements to be fulfilled*'.
- 'Theory' provides an explanation of the different legal concepts, additional information and a link to the relevant legal texts. For example, *Bioethics Law nr. 2004-800, August 6, 2004 (Article L1245-5 of the French Public Health Code); Ordinance no 2007-613, April 26, 2007; Decree no 2000-156, February 23, 2000.*
- 'Practice' provides information on the legal and/or administrative forms and procedural requirements. In the France- UK example, the researcher will have to ask for an authorization and will have to fill in a specific form that can directly be downloaded from the website. The procedure to be followed is described on the webpage.
- 'Issues' provides a series of questions and answers on common issues and provides references to articles and other relevant documentation. '*Many revisions of the law have been made, but there is no update of the cross references between the different parts of the regulation of import and export of biological samples for research. This regulation excludes gametes, gene and cell therapies. A new regulation is covering these activities since May 2008, but is only concerning therapeutic uses*'.

The latter three categories (Theory, Practice and Issues) are organized in the same systematic manner, that is, the legal issues are dealt with in the same consistent order. Each category will be divided according to the kind of HBM where relevant: blood, cells, tissues, serum, DNA. Furthermore, the international texts (including European texts) are provided on the main menu.

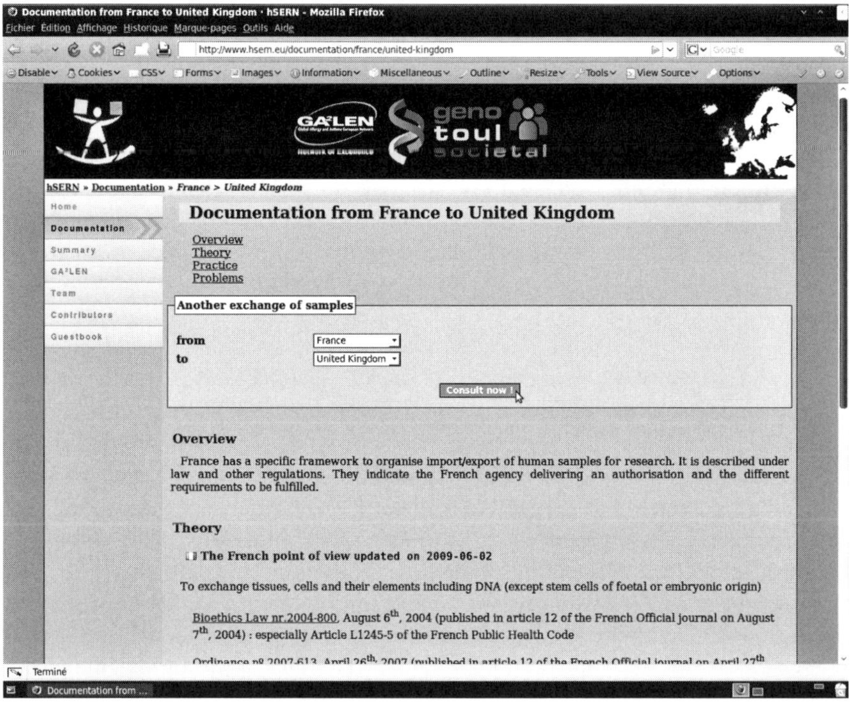

Figure 12.2 Example of hSERN webpage

Announcement of the adoption of new regulations

In the link marked 'News' hSERN will document any changes in the legal requirements and indicate the nature of the relevant text (whether it is law, regulation, or guidelines), its scope and the dates of adoption and entry into force if relevant. After analysis by the lawyers, the text will be included in the 'Theory' section and its practical effects will be described in the box 'Practice'.

What are Our Milestones?

Technical implementation

From a technical point of view, we must continuously update the content of the database. We must adapt the structure of the questions and refine the legal explanations in the tool in order to take into account the changes in the procedural requirements and the laws as well as to further develop the tool.

Relevant legal information validation

Access to the legal information in the various countries has to be regularly validated and updated. This will entail three requirements. Firstly, a strong network of lawyers needs to be identified as specialists in this topic. Secondly, the quality of the information provided has to be ensured. It depends notably on its regular updating. Thirdly, we have to develop mechanisms to ensure that the tool will meet the needs of the scientific community and that it provides an easy access.

Sustainability of the tool

For any tool, long-term sustainability is always an issue and it is necessary to identify and establish various incentives to ensure that such a tool can be maintained. The initial idea for this tool came from the GA^{2}LEN project, an EU project that has a limited time-frame and funding. There needs to be some way of rewarding the input of jurists in different countries who contribute their knowledge and expertise. The concept of micro-attribution (see, http://blogs.nature.com/ng/freeassociation/2008/03/microattribution_for_community_1.html) that has been proposed for scientists participating in maintaining databases could be one route to consider. Microcitation is the basis of microattribution of contribution to science. Microcitation is a way to incentivize public data deposition by extending the practice of citing journal articles to database entries and by providing quantitative citation for every unique author. hSERN can also be seen as one among several tools that can be used by 'biobankers' and biobank users. It needs to be developed according to the principles and conditions set up by relevant infrastructures such as the Pan-European Biobanking and Biomolecular Resources Research Infrastructure (BBMRI) and Public Population Project in Genomics (P^{3}G).

Conclusion and Future Developments

Managing such a web based tool is quite a challenge, because a website manager and a legal experts' network are needed in order to make it feasible. Time needs to be spent on validating, and updating legal information and also making it available in a user-friendly way for a given community of users. Contributors need to work together and to speak the same language. Such challenges have already been addressed for many other tools used by scientists, but the gap in language and cultural habits is wider between jurists and computer scientists than between biomedical scientists and computer scientists. The difficulty is the translation of the legal information into a practical and homogeneously structured web-based information system, in a comprehensive way and in a sustainable structure. Moreover, we will need to work to ensure this tool can to be continued once the GA^{2}LEN project has ended. It will be necessary to make the tool accessible beyond the GA^{2}LEN community and for the information to be broadly available for other research projects. This has already been envisaged within the BBMRI project. In addition, we will also have to develop a more specifically oriented tool for different types of samples as there is specific legislation for some types of HBM, such as embryonic stem cells and gametes. In order for the tool to be comprehensive other kinds of activities need to be addressed, such as the rules regarding the conservation and transport of samples .Finally, since medical data is often related to samples, it would be very useful to extend the database to the rules governing the exchange of such data.

Acknowledgements

The following European Union (EU)–funded projects supported this work: EU-FOOD-CT-2004-506378, the GA²LEN project (Global Allergy and Asthma European Network), EU Coordination Action FP6-LSHGCT-518148, PHOEBE (Promoting Harmonization of Epidemiological Biobanks in Europe) and EU FP7-212111 infrastructure project BBMRI (Pan-European Biobanking and Biomolecular Resources Research Infrastructure). The authors are also very grateful to Francine Kauffmann who provided very useful comments in order to orientate this tool towards questions that are faced by biomedical researchers.

Bibliography

Biobanking and Biomolecular Resources Research Infrastructure. 2008. [Online] Available at: www.bbmri.eu/ [accessed 3 April 2009].

Cambon-Thomsen A., Rial-Sebbag E. and Knoppers B.M. 2007. Trends in ethical and legal frameworks for the use of human biobanks. *European Respiratory Journal*, 30, 373-382.

European Society of Human Genetics Recommendation. 2003. Data storage and DNA banking for biomedical research: Technical, social and ethical issues. *European Journal of Human Genetics*, 11(Suppl 2), 8-10.

Ga²len project. [Online: Global Allergy and Asthma European Project]. Available at: http://www.ga2len.net/index.cfm?CFID=224703&CFTOKEN=47753371 [accessed 3 April 2009].

Kaye, J., Heeney, C., Hawkins N., de Vries, J. and Boddington, P. 2009. Data sharing in genomics – re-shaping scientific practice. *Nature Reviews Genetics*, published online: 24 March 2009.

Public Population Project in Genomics. [Online] Available at: www.p3gconsortium.org [accessed 30 May 2009].

UNESCO. 2003. *International Declaration on Human Genetic Data*. [Online] Available at: http://portal.unesco.org/en/ev.php-url_id=17720&url_do=do_topic&url_section=201.html [accessed 3 April 2009].

Zika, E., Schulte In den Bäumen, T., Kaye, J, Brand, A. and Ibarreta, D. 2008. Sample, data use and protection in biobanking in Europe: Legal issues. *Pharmacogenomics*, 9(6), 773-81.

Chapter 13

Biobanking Networks – What are the Governance Challenges?

Jane Kaye

Over the last 10 years we have seen considerable investment in new biobanks as well as a growing recognition of the importance of existing sample collections for secondary research purposes. Considerable effort is being put into ensuring that common standards are developed for new and existing biobanks in terms of standard operating procedures; the development of new methods and software for data and sample collection; and the development of template forms for informed consent and material transfer. The aim of standardization is to improve quality, increase efficiency and better facilitate research so that researchers can access, use, and compare data and samples from many biobanks. Many organizations are involved in the development of standards, which undergo a process of formulation and agreement; implementation; and then verification and updating. Examples are projects such as P³G[1] (population biobanks) and PHOEBE[2] (epidemiological biobanks and longitudinal cohort studies), as well as international organizations such as ISBER[3] (biological samples).

The next logical step from the development of common standards and procedures is the establishment of biobanking infrastructure that will lead to the development of networks of biobanks. The development of an infrastructure will enable the comparison of samples and information between biobanks and larger resources to be assembled. This will further stimulate the growth of research networks or 'networks within networks' that currently exist, as information will be more easily be able to be shared between individuals and research clusters, as well as the possibility to network biobanks. A network of biobanks would enable researchers to be able to access a number of biobanks at the same time through one portal rather than approaching each biobank individually for access. The first step in this process towards research infrastructure for medical research has been

1 Public Population Project in Genomics P³G Consortium, available at: http://www.p3gconsortium.org/ [accessed 19 April 2009].

2 Promoting Harmonisation of Epidemiological Biobanks, available at: http://www.phoebe-eu.org/eway/default.aspx [accessed 19 April 2009].

3 International Society for Biological and Environmental Repositories, available at: http://www.isber.org/ [accessed 19 April 2009].

started with a number of initiatives within Europe.[4] In the field of biobanking, the Biobanking and Biomolecular Resources Research Infrastructure (BBMRI)[5] aims to carry out the preliminary work for the development of an infrastructure to support the co-ordination of a network of biobanks within Europe. This infrastructure will facilitate and support wide-scale sharing of samples and data for all types of research.

The purpose of this chapter is to explore the issues raised by the development of biobanking networks, with a focus on genomic research, and the issues that this raises for governance. This chapter will be divided into three sections. The first will describe the vision for a pan-European biobanking infrastructure, which will lead to a biobanking network. The second will describe the technological and policy changes that have occurred in the field that provide the necessary conditions to allow and support large biobanking infrastructure and networks. The third section will focus on the challenge to basic principles that such initiatives create, particularly to informed consent, identifiability, and the protection of privacy. The final section will discuss what current challenges exist to the development of appropriate governance mechanisms for biobanking networks.

The Vision of Pan-European Biobanking Infrastructure

The purpose of the BBMRI is to develop 'pan-European biobanking infrastructure to manage human samples and data which supports world-class biomedical research and underpins the development of personalised medicine for European citizens' (Yuille 2009). The aim is to include already existing and de novo collections, and to build upon existing expertise and technologies within Europe. The European Strategy Forum on Research Infrastructures (ESFRI), an intergovernmental panel that identified these aims and recommended a structure such as BBMRI, anticipated that 'major synergism, gain of statistical power and economy of scale will be achieved by interlinking, standardizing and harmonizing – sometimes even just cross-referencing – a large variety of well qualified, up-to date, existing and de novo national resources' (Yuille 2008).

The purpose of pan-European biobanking infrastructure is to allow the sharing of samples and information by researchers to ensure that resources can be used for secondary purposes. This means that while resource management may be carried out by one biobank, the European network will allow samples and data to be managed, stored and curated using similar methods to make access by researchers easier. This management infrastructure includes 'the robots and freezers for sample

4 EATRIS is a European network of new research centres translating basic discoveries into clinical interventions in major diseases. ECRIN is a European network of clinical research centres, clinical trials and biotherapy facilities for therapeutic innovations.

5 Biobanking and Biomolecular Resources Research Infrastructure website, available at: http://www.bbmri.eu/ [accessed 19 April 2009].

management; the computers, servers and software for data management; and the data networks and administrative systems for access management'.[6] The vision is that by using a 'hub and spoke' model with grid computing technology, pan-European infrastructure will enable sample and data sharing.

It is envisaged that a pan-European biobanking infrastructure will have a legal personality that will be comprised ultimately of EU member states, but each member state will determine the way that biobanking is organized within their jurisdiction. The aim is that biobanks, biomolecular resources and technology centres will be members of the infrastructure and connected to their specific hubs in each country. This kind of architecture is currently needed because different biobanks have been established to support different kinds of research with different requirements and needs. Thus, the ambition for a pan-European biobanking infrastructure is to overcome fragmentation in the European research area by enabling wider cooperation in resource management while still allowing for the diversity of research needs.

Changes in the Field of Genomics

The vision for a European biobanking infrastructure is only possible because of the rapid progress in the use and application of new technologies within medical research. In the field of genomics the most significant of these are in sequencing and computing that has led to the new field of bioinformatics. New sequencing technologies enable scientists to move from a focus on single genes to whole genome sequences, and to embrace genomics as with a wider scope. The field of bioinformatics has allowed datasets to be integrated, compared and organized to create virtual libraries or research resources, such as ENSEMBL,[7] HapMap,[8] the 1000 Genomes Project,[9] the European Genotype Archive[10] and dbGaP.[11] The richness of the data now available, combined with the powerful bioinformatics tools has led to data mining of large datasets to explore the associations between existing data points. The advances in sequencing technology and bioinformatics have in turn had an effect on scientific practice, which 'has become increasingly interdisciplinary, with the rapid formation of flexible and dynamic research collaborations around the world' (Kaye et al. 2009). There is the potential for a large number of people to be involved in the distribution and creation of data.

6 Personal communication with Martin Yuille, University of Manchester, UK, 20 May 2009.

7 http://www.ensembl.org/index.html [accessed 19 April 2009].

8 http://www.hapmap.org/ [accessed 19 April 2009].

9 http://www.1000genomes.org/page.php [accessed 19 April 2009].

10 http://www.ebi.ac.uk/ega/page.php [accessed 19 April 2009].

11 http://www.ncbi.nlm.nih.gov/sites/entrez?Db=gap [accessed 19 April 2009].

The development of the scientific agenda has also been intertwined with these technological developments. The subtle interactions responsible for common diseases can only understood by having large representative samples from different populations (Burton and Hansell et al. 2009). 'The discovery of critical genes and pathways as well as the follow up analyses of the impact and significance of these will critically depend on biobanks, large collections of well-documented, up-to-date epidemiological, clinical and biological information and accompanying material from large numbers of patients and healthy persons representing the general population. Biobanks are a key resource in unravelling the association between genetic background, life style and environmental risk factors for various diseases and their trait components' (BBMRI website 2008). This has led to the establishment of large population biobanks but it also requires that metadata, samples and health information are characterized in a detailed and organized manner.

These changes have been encouraged and facilitated by funders who have provided the resources to develop the technology and in doing so have spurred changes in scientific practice. Over the last 13 years we have seen funding bodies on both sides of the Atlantic taking an active role in supporting and directing new open access policies, such as the Bermuda Agreement in 1996, which was followed by the Fort Lauderdale Agreement in 2003 and most recently, the Toronto meeting in May 2009. These documents together set out the key principles which now dominate thinking and practice regarding open access to genome sequence data in North America and the UK. Funders also have been instrumental in initiating data-generating infrastructures, as well as developing open access policies in the area of publishing. The rationale behind open access policies is to ensure that advances in science will be accelerated by opening up data and knowledge to all. This makes the maximum use of resources that have been funded by the public purse. These policies have only been possible because of an active dialogue and co-operation between the scientific community and the funding institutions. Funders have been able to push these policies forward because they have had the support of the scientific community but also because they have the resources to fund such ventures. The development of a global biobanking infrastructure will also be dependent upon large-scale funding and the co-ordination and synchronisation of a funding programme at an international level.

Due to these policies we have a situation where there are 'networks within networks' enabling researchers to explore multiple datasets, making connections and associations between them in a way that has never been possible before. With the click of a mouse, researchers can seamlessly move between datasets held in different institutions while at the same time crossing national jurisdictional boundaries in virtual space. The effect on practice has been to make research less focused on data collection and establishment of your own dataset, that you can mine exclusively for many years, to a type of research which is less hypothesis focused but can be based on making associations between datasets (data-mining), enabling researchers 'to stand on the shoulders of giants'. Research data and access to it, is now possible on a scale that was never envisaged before, as datasets (in contrast to samples) are an infinite resource. The secondary users of the data are far removed from the researchers who carried out the collection of the samples and data, as well as the research participants who are the source of the samples and information. The BBMRI vision is only possible because of these significant changes in the way that the science is being carried out and these technological developments that produce large amounts of data that can be shared easily around the globe.

Testing Fundamental Principles

The development of a biobanking networks test many of the principles that have applied to medical research. In this chapter I will focus on the key issue of how to protect research participants' interests. A biobanking network challenges the efficacy of informed consent; our ability to protect privacy; and current governance structures.

Informed consent

Informed consent and the right to control how personal information is used are foundation principles in legal documents on medical research and in human rights law. The most widely accepted definition of informed consent is found in the WMA Declaration of Helsinki:

> In medical research involving competent human subjects, each potential subject must be adequately informed of the aims, methods, sources of funding, any possible conflicts of interest, institutional affiliations of the researcher, the anticipated benefits and potential risks of the study and the discomfort it may entail, and any other relevant aspects of the study. The potential subject must be informed of the right to refuse to participate in the study or to withdraw consent to participate at any time without reprisal. (WMA 2008: Art. 24)

In addition, for the consent to be valid the person giving the consent must be competent and the consent must be given voluntarily and without coercion.

The key requirements of this formulation of informed consent that are problematic for biobank infrastructure that facilitate data sharing are: that consent must be obtained before the research starts; that all the researchers and research uses must be stipulated at the time the consent is obtained; and that individuals must be able to withdraw consent at any time. It is virtually impossible to achieve informed consent as envisaged in the Declaration of Helsinki in a networked, global world, where data sharing is fast becoming a norm for continued funding. One of the perceived benefits of open access is that the data can be used by a number of researchers for many different purposes at the same time and that there is the capacity to be able to 'play', 'get your hands dirty' and explore the raw data. Any restrictions on that has the potential to lessen the benefits and the speed that these benefits can be realized.

In the world of biobanking, where many of these issues have been grappled with for some time, 'broad consent' has been seen as a practical solution to these consent issues raised by research resources. With broad consent, an individual gives consent to genomic research, on the understanding that the data will be used for many future uses, rather than for one specific use. As part of a broad consent individuals may give consent to be re-contacted for other research projects that may require additional information or samples. However, currently, with a broad consent research participants do not have to give additional consent for new research uses of their information. They will have no control over their information once they have given consent to the collection of the information. This policy, which has developed out of practical constraints of having to re-contact thousands of participants in population biobanks, is still controversial and hotly debated. With the creation of biobanking networks, the consent issues are multiplied with the ability of researchers to access and compare a number of biobanks simultaneously rather than just having access to one biobank.

This raises the question whether a broad consent is adequate for data sharing that is carried out through a biobank network. Much of this rests on the idea of autonomy that we are seeking to protect. For example, some would regard that for an individual to be able to exercise their autonomy and decide upon their involvement in medical research, they should know all the details of the research uses (Salvaterra et al. 2008, Caulfield 2007, Brekke and Sirnes 2006, Kaye 2004). However, in the case of a broad consent for biobanking, some have argued that it is sufficient to know the 'rules of the game' (Nõmper 2005) – rather than all the details of the research that will be carried out on the biobank. On this basis, if an individual has agreed to put their information in a biobank for many possible uses and they know that this will be used as a resource, having further details about this being part of a larger research network should not be problematic. However, this requires a great deal of trust on behalf of participants and does not allow the individual to decide on the research that they are involved in.

Having a broad consent for the data access carried out within a biobanking network starts to test the effectiveness of broad consent as a reflection of individual autonomy. I would argue that by giving broad consent an individual is effectively giving a broad consent to inclusion in a biobank but then gives 'consent to governance' – or consent for others to decide upon their behalf. Within Europe, in cases where the consent is in dispute or it is unclear whether the new research is in the scope of the original consent, referral maybe made to an independent body, that is often a research ethics committee. This is embodied in a number of legal documents but has largely been seen as an exception to the requirements for informed consent and has only been deemed appropriate for use in certain situations. Now when data sharing is becoming more routine, 'consent to governance' is becoming the norm rather than the exception. In biobanking networks, without technological mechanisms for individuals to control how their data is used, the only possible way forward is to establish independent bodies that can make decisions on behalf of research participants. The way that this might be achieved has not yet been decided and is still open to discussion.

Protecting Privacy

The world has changed since the completion of the Human Genome Project in 2003. More and more data is made available through the web and is able to be accessed by 'the world at large', recombined and then once again deposited on the web for another cycle of reiteration. The information that is available can be highly detailed, derived from different sources, and can easily be matched all in the comfort of your own space. The internet has the effect of bringing information to you, within reach of your finger tips. Our principle of openness and the fact that we can put information from our space into the wider space can sometimes blind us to the implications of our actions. Once information is on the web, we no longer have control over its use or further dissemination, or currently the ability to claw it back.

In the sphere of medical research, where a paramount concern is safeguarding the interests of research participants, important principles are maintaining the privacy of research participants and the assumption that data must be keep confidential. For this reason, the Wellcome Trust and the National Institutes of Health decided to remove SNP data created using GWAS methodology from the publicly accessible databases of the Wellcome Trust Case Control Consortium (WTCCC)[12] and the Genetic Association Information Network (GAIN Project).[13] This action was precipitated by a paper by Homer and others (2008) that established through a statistical analysis that an individual could be identified in aggregate data, because of the unique data points that exist on an individual genome. Cases of using the internet to identify individuals are becoming more common, such as the famous case of the boy who tracked his sperm donor father down (Motluk 2005) and the recent use of publicly available datasets to identify individuals (Gitschier 2009). This is not surprising as DNA is a unique identifier and it is impossible to completely anonymize a sample, as relatively small numbers of SNPs can be used to identify individuals if a third party has SNP information on the individual concerned (Nyholt et al. 2008).

In a world where digital information is increasingly available on the web the possibility of maintaining participants' privacy is increasingly challenged. In the case of many small projects, it may still be possible to use security devices such as coding and firewalls to keep sequence information non-identifiable to researchers. However, this position becomes increasingly tenuous in the case of longitudinal

12 www.wtccc.org.uk [accessed 24 May 2009].

13 http://www.genome.gov/19518664 [accessed 24 May 2009].

projects such as population biobanks and if data can be obtained outside of the sphere of medical research where there are appropriate controls and protections. It is access by people other than researchers that is of concern, as people can increasingly have access to their own genome through companies such as 23andMe,[14] Navigenics[15] and deCODEME.[16] Therefore, the development of biobanking networks must have appropriate controls to protect individual information when data is shared both within the research community as well as ensuring that it is protected from those outside. The potential for privacy infringements and third party researchers being able to identify individuals, increases with the comparison and integration of different datasets held within biobanks. However techniques such as 'virtualization' will eliminate the need for a third party to access raw data, as researchers will be able to obtain the information that they need from metadata alone, without having access to the raw sequence data.[17]

Governance Challenges

One of the significant challenges for the development of a biobanking network is the different legal and governance frameworks that exist across Europe, but also across the globe for medical research. Our current legal frameworks and regulatory structures are nationally based and vary between countries, but our research activity is global. This means that we have different legal instruments that apply in different countries. For example, with internet searching it is possible to cross many jurisdictional boundaries by using different web sites that are registered in different countries. The transfer of samples from biobanks to researchers or to institutions through a biobanking infrastructure network raises similar issues. In addition, our research governance systems are nationally based with research ethics committees being the primary body that approves research. This situation creates problems for biobank networks.

Different legal requirements

Within Europe, we are moving towards greater uniformity with common legal instruments to protect privacy such as the Directive 45/46/EC and case law on the Art. 8 right to life from the European Court of Human Rights. These would have general application to data within the medical research context, but there is no legal instrument in Europe that deals specifically with this issue. Even though we have common legal instruments at a European level, there are still considerable

14 https://www.23andme.com/ [accessed 24 May 2009].

15 www.navigenics.com/ [accessed 24 May 2009].

16 www.decodeme.com/ [accessed 24 May 2009].

17 Personal communication with Martin Yuille, University of Manchester, UK, 20 May 2009.

differences between the UK and other member states as a 'margin of appreciation' is allowed for the implementation of European Union law into national law. For samples, the difference in the law between jurisdictions is even more acute, as there is no common overarching European legal instrument. This means that there are different legal requirements for samples and data between jurisdictions.

At an international level there are documents that lay down the principles for medical research, but many of these are not mandatory and are dependent on national governments signing up to them. For example, in the case of the Convention on Human Rights and Biomedicine 1997, the key document within Europe on medical research, the UK has not signed this document because of the provision on therapeutic cloning. While international documents can provide a baseline of general principles in medical research, they are not the best way to regulate emerging areas – until the technology is well established and it is demonstrated that there is sufficient need for regulation. These higher level legal instruments provide a framework, but do not, and cannot, provide all of the details needed for the development of standards and ethical principles that are needed to support biobank networks.

In the case of new areas of practice, standards or guidelines are often developed by professional communities. These provide direction and yet can be updated easily to accommodate new developments. These standards are not at the level of law and could always be overturned in a conflict with national law. However they present one solution to the problem of global harmonization and are a means of developing ethical standards for practice that can apply when the legal standards are not evident, or are too general to have sufficient application. Agreed standards and principles are essential for the development of research networks in order to provide a framework for research activity and the sharing of samples and data.

The lack of appropriate bodies

If we accept that individuals must give 'consent to governance' for the secondary use of personal information then we need to ensure that there are bodies which can make decisions and act on behalf of individuals. Research ethics committees could have an important function in monitoring data access and protecting the privacy interests of participants. It has become the norm in most countries, which is sometimes enshrined in national law, but not always, that approval by a committee is required for all research that is carried out on human beings. Such committees are one of the key gate keepers in the ethical review process, as they have the power to allow research to proceed or not. The purpose of these committees is to ensure that the interests of research participants are protected and that scientific research is carried out in an ethical manner according to accepted scientific practice. Therefore, it would appear that they are in a key position to oversee the ethical issues raised by research projects, but it is not clear that they are appropriate for biobanking networks.

The benefits of using research ethics committees are that their authority is recognized and respected. However, while these committees are independent, they consist of volunteer professionals and lay people, who may, or may not, have ethical training. It not evident that many research ethics committees would have the expertise to understand the privacy concerns associated with sharing samples and data from many biobanks through a biobanking network and if they would be able to carry out a privacy assessment to establish the disclosure risks. In addition, the decisions of research ethics committees are not uniform and can vary between regions in countries and between countries. Decision making about data access to a hub and spoke biobank network needs to be uniform, transparent and not dependent upon where the researcher files an application for access. Therefore, in order for ethics committees to take on this role, they would have to develop a system of uniform, transparent and accountable decision-making that was recognized across Europe. The ideal for researchers is to have access approval for one biobank based in one country that can also apply to another biobank that was part of the same network.

New governance models have been established to regulate access to datasets of detailed patient information, as well as sequence information that has been created by data-generating projects developed using a genome-wide association studies (GWAS) approach. 'Data access committees' have been established for the WTCCC and GAIN projects that generate genome-wide data, to determine who should have access to data and on what grounds. These projects are virtual repositories of information from a number of researchers and research projects. The basis for decision-making by the data access committee involves establishing whether a scientist is a '*bona fide* researcher'. However, the bases for assessing this are still in the process of being developed and are not uniform or publicly known. These committees are in addition to the research approval that must be obtained from a research ethics committee as data access committee approval is necessary only for access to the dataset. In the case of data-generating projects, each project has its own data access committee, which a researcher must apply to for each new research project. This means researchers must get a new approval from each project they wish to access. If at some point there was a proposal to integrate the different projects, then this policy may have to be re-thought. One of the privacy issues is that there is no one body that is looking at disclosure when all of the genomic datasets are analysed together.

The development of a biobanking network raises similar issues. It is evident from the experience of GWAS projects that new independent access committees were regarded as necessary, and additional to the oversight of research ethics committees. The issue for a hub and spoke network is whether access is determined by individual biobanks, or whether one approval to a researcher gives access to the network as a whole and all the biobanks within it. The aim is to develop a governance system that is not too onerous while regulating the right kind of activity to ensure that privacy is safeguarded and that things do not occur that might jeopardise the research and public trust. As well as being appropriate, such a system must be independent, transparent and accountable in its decision-making and in the procedures that are established to make decisions. Researchers must know the reasons for decisions; which body has made them; and the procedures that must be followed to appeal decisions which they feel are unjust. Such a system must also be able to transcend national boundaries but also be in compliance with national law. The GeneLibrary Ireland study design proposes that ethics approval must be obtained as usual and that additional approval must be obtained by the Governance and Ethics Committee established for GeneLibrary Ireland (GeneLibrary Ireland 2009). However, this does not deal with the issue of having a number of networks that are joined together internationally via the internet, as researchers still have to apply to each project in each country.

Conclusion

The purpose of this chapter has been to sketch out some of the issues that are raised by the development of a biobanking network within Europe. This raises a number of challenges for informed consent and privacy protection but most importantly, how access to biobanks should be governed. In the area of informed consent, the development of a biobanking network will accelerate a change that is already occurring, from informed consent to broad consent coupled with 'consent to governance'. However, if we are going to rely on this as a basis for wide-scale data sharing for research purposes, we need to make sure that the bodies that are making approvals on the basis of individuals are accountable, independent and representative of research participants. As they are currently constructed, research ethics committees and data access committees do not have the appropriate credentials to fulfil this role. It is also not evident that they have the capacity or mandate to make assessments of privacy risks. This is particularly so in the case of wide-scale access to data and samples through a biobanking network.

A biobanking infrastructure for Europe is still in its infancy and plans are still in the process of being developed. Ultimately, this will lead to a biobanking network, which will strengthen the many research networks that exist but also push this to a new level of data sharing. A fundamental question is how to govern access to data and samples held within biobanks in a hub and spoke infrastructure, so that research participants' privacy is protected and yet research can proceed without undue hindrance. The purpose of this chapter has been to highlight some of these issues to provide a basis for thinking further about the development of appropriate global governance mechanisms that can maintain this careful balance.

Acknowledgements

The research for this chapter has been funded by the Wellcome Trust Grant 081407/Z/06/Z and by P^{3}G as part of the Ethox Centre's P^{3}G Core on Data Sharing (2009).

Bibliography

Brekke, O.A. and Sirnes, T. 2006. Population biobanks: The ethical gravity of informed consent. *Biosocieties*, 1(4), 385-398.

Burton, P.R. and Hansell A.L. et al. 2009. Size matters: Just how big is BIG? Quantifying realistic sample size requirements for human genome epidemiology. *International Journal of Epidemiology*, 38(1), 263-273.

Caulfield, T. 2007. Biobanks and blanket consent: The proper place of the public good and public perception rationales. *King's Law Journal*, 18, 209-226.

Gitschier J. 2009. Inferential genotyping of Y chromosomes in latter-day saints founders and comparison to Utah samples in the HapMap Project, *American Journal of Human Genetics*, 84, 251-258.

Kaye, J., Heeney C., Hawkins, N., de Vries J. and Boddington, P. 2009. Data sharing in genomics – Re-shaping scientific practice. *Nature Review Genetics*, 10, 331-335.

Molecular Medicine Ireland. 2009. *The GeneLibrary Ireland Design Phase Report*, Health Research Board and the Health and Social Care Research and Development Office.

Motluk, A. 2005. Anonymous sperm donor traced on internet. *New Science*, 2524, 6.

Nõmper, A. 2005. *Open consent – a new form of informed consent for population genetic databases*. PhD Thesis. University of Tartu, Estonia.

Nyholt, D.R., Yu, C.E. and Visscher, P.M. 2008. On Jim Watson's APOE status: Genetic information is hard to hide. *European Journal of Human Genetics*, 17(2), 147-9.

Salvaterra E., Lecchi L., Giovanelli, S., Butti, B. et al. 2008. Banking together: A unified model of informed consent for biobanking. *EMBO Reports*, 9(4), 313.

The Bermuda Principles. 1996. First International Strategy Meeting on Human Genome Sequencing, Bermuda, 25-28 February 1996. Available at: http://www.ornl.gov/sci/techresources/Human_Genome/research/bermuda.shtml#1 [accessed 19 May 2009].

Wellcome Trust. 2003. *Sharing Data from Large-scale Biological Research Projects: A System of Tripartite Responsibility*. Report of a meeting organized by the Wellcome Trust, Fort Lauderdale, 14-15 January 2003. Available at: http://www.genome.gov/Pages/Research/WellcomeReport0303.pdf [accessed 19 May 2009].

World Medical Association. 2008. *Declaration of Helsinki Ethical Principles for Medical Research Involving Human Subjects*. Adopted at the 59th WMA General Assembly, Seoul, October 2008. [Online] Available at: http://www.wma.net/e/policy/b3.htm [accessed 19 May 2009].

Yuille, M. et al. 2008. Biobanking for Europe. *Briefings in Bioinformatics*, 9(1), 14-24.

Yuille, M. 2009. *Update on the Operational Concept of BBMRI*. Powerpoint presentation at the Joint Meeting of BBMRI Participants and Associated Members, Brussels, 25 March 2009.

PART 4
Governing Bodies

Chapter 14
Potential Conflicts in Governance Mechanisms used in Population Biobanks

Karine Bédard, Susan Wallace, Stephanie Lazor
and Bartha Maria Knoppers

Governance mechanisms are used by biobanks to ensure that the public trusts, and wishes to support and participate in, the research of a population biobank. But when, in what contexts, and in what combinations should these mechanisms be used in order to ensure a biobank acts in the public's interest? This chapter outlines research conducted at the request of the Québec Ministère de la Santé et des Services Sociaux (MSSS) in Canada. The MSSS was interested in understanding what elements would be necessary for an adequate and effective governance structure for the CARTaGENE biobank in Quebec that was funded in 2007. They also wanted information on how the Institute for Populations, Ethics and Governance (IPEG), created to provide oversight of CARTaGENE, should be designed, and if it would be the appropriate mechanism for this role. In order to investigate these questions, the governance mechanisms used by member biobanks of the Public Population Project in Genomics (P^3G) were analysed. This chapter documents these findings and the reason why the MSSS decided to disband the IPEG and rely on existing governance mechanisms to oversee the CARTaGENE biobank.

The Changing Context of Genomic Research

Research practices have changed over the years. Rather than working alone in their own laboratories, researchers now share data and samples internationally. Reflecting this, the number of biobanks or population resources, defined as large-scale collections of biological materials and their associated data (Cambon-Thomsen 2004), has increased exponentially. These resources are now seen as a prerequisite by many scientists who want to conduct studies on complex diseases and genetic epidemiology. As Seminara et al. (2007: 1) have written, these resources help researchers '… to identify genes underlying complex common diseases, to assess associations between genetic variants and disease susceptibility, and to examine potential gene-environment interactions'.

These biobanks, however, must deal with a different set of ethical issues than traditional clinical research. These include participants that are disinterested citizens as opposed to those with a specific disease or condition; the need for a broad, rather than specific, consent; and the confidentiality and privacy concerns that come with large datasets (Greely 2007, Wallace et al. 2008). In order to ensure the trust of the participants and the public, these biobanks must be governed in such a way as to take into account the differences between longitudinal and traditional clinical research.

Governance frameworks have further been described as, '... the agreements, procedures, conventions or policies that define who gets power, how decisions are taken and how accountability is rendered' (Institute on Governance 2003:1). Governance, as it relates to these resources, has been defined as, '[t]he process of policy orientation and management that guides and regulates research under ethical and scientific norms so that the results can be used for the benefit and improvement of the health of the population' (P^3G Observatory 2008). As Deschênes and Sallée (2005:40) assert, 'It is not enough to ask a whole population for unquestioning trust; one must put in place good governance structures and mechanisms to ensure that the projects follow through with their promises to participants'.

The CARTaGENE Biobank

CARTaGENE is a public resource with the aim of collecting materials to assist researchers to accelerate health and genomics research. Based in the province of Québec in Canada, the resource consists of a databank containing health information and a biobank containing blood and urine samples, collected from individuals in Québec. Funded in 2007 by Genome Canada and Genome Québec (Genome Canada 2007), CARTaGENE researchers completed the pilot phase of the project in March 2008, during which they recruited 223 people, testing their methods, tools and instruments for patient recruitment and data and sample (blood and urine) collection procedures. Phase A of the project, when 20,000 people between the ages of 40 and 69 will be recruited, is expected to begin in 2009 (CARTaGENE 2009). In a subsequent phase, another 30,000 individuals will be recruited. The resulting resource of data and biological samples will be accessible to local and international researchers conducting studies on risk factors (genetic, environmental and social), including the health status of the population. As a result of this research, it is hoped that future results from these studies will facilitate the implementation of better clinical diagnostic, treatment and prevention tools.

CARTaGENE is also a founding member of the Public Population Project in Genomics (P^3G) Consortium (P^3G 2009). P^3G has the mandate '... to create, harmonize and share methods, tools and information so as to enhance the design of emerging biobanks and to promote compatibility – between studies – of data (e.g. socio-economic and clinical), samples, and supporting infrastructure (e.g. sample and data-management systems)' (Knoppers et al. 2008: 664). As a P^3G member, CARTaGENE can harmonize, where appropriate, its methodology with other member biobanks in order to increase the pool of harmonized data and samples accessible for researchers and thereby achieve the necessary statistical power.

One important aspect of the development of CARTaGENE has been to determine its governance framework. As with other resources, the CARTaGENE project needed to put in place the bodies, committees, staffing, policies and processes necessary for it to be able to carry out its duties. In the early stages of CARTaGENE, one specific question asked was whether it should have an independent governance committee, modelled on that of UK Biobank, in addition to internal governance committees (such as a Board of Directors or Executive Committee). UK Biobank was one of the first large-scale population-based resources (Palmer 2007, Deschênes and Sallée 2005, Tutton 2004). One part of its governance framework is the UK Biobank Ethics and Governance Council (EGC). The funders of UK Biobank created the EGC as an independent oversight body in response to public consultations held in the early stages of the creation of UK Biobank.

Stakeholders had raised questions about the future uses of the data and samples that would be collected and about who would have access; abuse was raised as a potential issue. The stakeholders consulted agreed that '… the establishment of a truly independent oversight body … [was] … paramount to the success and safety of the project' (Opinion Leader Research 2003: 19). As a result, EGC's mandate is '[t]o act as an independent guardian of the UK Biobank Ethics and Governance Framework (EGF) and advise on its revision; to monitor and report publicly on the conformity of the UK Biobank project with the EGF; to advise more generally on the interests of research participants and the general public in relation to UK Biobank' (EGC 2009). It oversees, yet is independent from, UK Biobank, making it different from other traditional ethics committees (Deschênes and Sallée 2005).

In 2003, the non-profit Institute for Populations, Ethics and Governance (IPEG) was formed. It was given '… the mandate of managing the CARTaGENE project and other biobanks in Québec that wish to be so governed. IPEG will ensure that the CARTaGENE research platform receives ethical and scientific approval before its launch and is used in the best interest of the public' (Deschênes and Sallée 2005: 42). According to Québec law, local research ethics boards will review the ethics of projects that wish to use the CARTaGENE resource. IPEG, on the other hand, would have created policies for the use of the resource's data and samples, as well as ensuring that approved uses were in conformity with the mission of CARTaGENE.

Governance Mechanisms Used by Population Resources

During the consultation that preceded the funding of CARTaGENE, questions had been raised about the structure and role of IPEG in relation to CARTaGENE. In 2006, when CARTaGENE was finally launched, the Québec Ministère de la Santé et des Services Sociaux (MSSS) approached the CRDP to do additional research into the governance needs for CARTaGENE, as well as into the specific structure and remit of the proposed oversight body, IPEG. The MSSS wished to ensure that CARTaGENE was adequately governed; it also did not want to impose a committee structure that might stifle development or hinder efficiency.

Methodology and results

The CRDP embarked on research to identify the governance frameworks and mechanisms used by existing population resources. They surveyed the internet-based and other publicly available materials of 18 biobanks that were members of the P³G Consortium. This research was limited to information available in Spanish, French and English.

First, a framework of elements that were common to all the biobanks was identified. This common framework consists of six 'elements' or sets of governance procedures: (1) scientific evaluation; (2) ethics evaluation; (3) data protection and public health laws, and laws on statistics; (4) bio-security standards for laboratories; (5) guidelines for research with human participants; and (6) professional guidelines. When these common elements were set aside, the differences between the biobanks became clearer. The analysis showed that some resources were located in a country with a biobank-specific law, such as Estonia, Spain and Sweden, which necessarily influenced that biobank's governance framework (Nys and Fobelets 2008, Gibbons et al. 2005). Other resources, such as UK Biobank, Generation Scotland, and the Joondalup Family Health Study in Australia had specific governing bodies to oversee their activities (EGC 2009, Generation Scotland 2009, JFHS 2006). Yet others relied on the existing laws, guidelines and other cultural norms in their country for their governance framework (Wallace et al. 2008). Together these external elements influenced the way in which the resources had shaped their governance framework.

Researchers also collected information on the governing bodies of these resources and looked for similarities and differences in them in four specific areas: mandate, membership, operation (that is, the way in which the resource's governance body functioned) and financing. Governance bodies examined in this research were categorized as either: (1) a body or committee specifically created to govern the resource; or (2) other more general bodies or committees that were not specifically created to govern the resource but which fall under the common framework and play some oversight role for the resource. When examining the four specific internal mechanisms, many differences were seen in the way in which resources' governing bodies were constituted and how they operated.

Table 14.1 Key similarities and differences between biobanks' governance bodies

Categories	Elements common to all resources	Elements that differed between resources
Mandate/Role	Ensuring that: The resource conforms to internal management rules The resource conforms to ethical, legal and scientific standards Results are shared with the population The confidentiality and privacy of the participants is respected	Carrying out audits of the databases and/or the data gathered Organizing consultations with the public Ability to sanction the resource if they do not follow rules and regulations, including: • Publicly denouncing the resource • Preventing access to the resource • Closing the resource
Membership: Expertise	Including experts in ethics, law and science or medicine	Including experts in philosophy, sociology and/or epidemiology Including biobank participants or members of the population
Membership: Source	Recruiting members who are citizens of the country where the resource is located	Recruiting/consulting with international members/experts
Operations	Ensuring that members are all independent of the resource Appointment of members by people who are independent of the resource	Permitting resource personnel to attend meetings, but without the right to vote Providing clear term limits for members Indicating the frequency of meetings Detailing how activities are reported
Funding	Receiving financing from non-profit public agencies or governmental organizations (no private financing)	Receiving financing from the same organization that funds the resource, but from different funding streams within that organization Receiving financing directly from the resource, but organizing the committee as an independent body

Mandate All the bodies studied mentioned in their description that they had a duty to ensure that the resource conformed to internal management policies and applicable ethical and legal requirements. The majority summarized their mandate, but some provided additional details regarding their audit functions or their role in organizing public consultations. Some governing bodies had the ability the refuse access to researchers, or for some types of projects, if they did not conform to the mission of the resource.

Membership Among the projects that described the composition of their governing body, all mentioned the importance of having experts in law, ethics and science. However, the majority were vague on who was a 'scientific expert'. In addition, many did not mention the inclusion of a member of the public or representative of the population being studied. The majority did not detail from where members were recruited (that is, whether they had to be a resident of the region or country in which the resource was located). Certain resources noted that they invited members from the international community, in order to broaden the expertise or to limit potential conflicts of interest.

Operations Few details on the operation of the governing body were provided. All mentioned the desire to have members who were independent of the resource and who were chosen by people without a financial or other interest in the resource. But little information was given on terms of office, the frequency of meetings or public reporting/disclosure policies.

Funding All the governing bodies were described as being financially independent of the resource over which they had oversight. But 'independence' was defined in different ways. Some said they were independent because their financing came from an organization other than the resource. Some indicated that they were funded by government agencies using public money and thus were independent, but did not specify whether the government agency was the same as the one that funded the biobank. Others were independent because the funds given to the governing body were separate in some way from the funds given to the resource, even though the source of the funds was the same (for example, funds coming from two different departments in the same agency). And finally, some were independent because the governing body was organized independently from the resource, even though the funds for both came from the same organization.

Issues raised by the research

When examining these four elements together with the common framework and other governance mechanisms that influence population-based resources, the research showed several areas in which there is a potential for confusion, conflict and redundancy in the governance frameworks used by a resource or biobank.

First, it showed that there is a potential for conflict between the governance body of the biobank and the external science and ethics review committees that granted approval for the use of the biobank itself. This was the most striking issue among the biobanks. Once the population biobank is approved by an independent review body, such as an ethics committee, what further role, if any, should it have? Should general oversight responsibilities be passed to the specific governance body of the biobank or should the ethics committee also continue monitoring the biobank's progress and processes? This brings into question the membership of the ethics committee and whether it has the expertise, competence and support necessary to carry out this function. It also questions whether or not ethics committees should relinquish their monitoring role over the biobank, something that might be dependent on local laws and regulations regarding an institution's responsibilities for their researchers and projects. There is the further concern of unnecessary redundancy, if both the governance body and the review committees oversee the same aspects of the biobank.

A second issue was potential conflicts that may be found between the mandates of ethics review committees and those committees responsible for reviewing requests from researchers for access to the biobank's data and samples. The systematic collection of data and samples over time for future research is the raison d'être of population biobanking projects and, for the majority, requests for access by external researchers working on specific diseases are expected and welcomed. However, biobanks must ensure that researchers requesting data and samples will use them for scientifically and ethically valid research that falls within the objectives of the biobank. Biobank participants must trust that those reviewing and approving these requests will act in their best interests and in those of society. Therefore, requests must be considered carefully so that data and samples are used efficiently, effectively and wisely. Who then should take on the responsibility of reviewing these requests – an internal data access committee or an external ethics review committee?

The membership of governing bodies was also examined. Considering that all of the biobanks surveyed are involved in genetic epidemiology research, few mentioned the importance of having members who were experts in genetic epidemiology, population genomics and/or biostatistics. All mentioned having a physician as a member; however, there is a question as to whether this is sufficient considering the specificity of these population biobanks. Equally absent were experts in data security and access systems, as were 'lay' members, as noted earlier. Guidance is available on ethics review committee membership (TCPS 1998), but little discussion has been had on which areas of expertise, in relation to population biobanks, should be represented in governing bodies or whether such expertise should be added to the committee in an ad-hoc manner.

Finally, an issue that may hinder public trust in biobanks relates to the independence of the governing body. A government may wish to have an independent body overseeing a government-funded project, yet if that body is also funded by that government, will that body be sufficiently independent to conduct its work? Will it be able to question its own funders if problems arise with the biobank? In the case of UK Biobank, the EGC can publicly denounce UK Biobank's practices if they feel it is necessary, but they cannot actually force UK Biobank to cease those practices. Whether this power is sufficient is unknown; the EGC has not yet reported on having to exercise this power. However, it does raise the question of whether this arrangement is sufficient to adequately protect the needs of the participants in UK Biobank (Tutton, Kaye and Hoeyer 2004). Third-party oversight funded by groups not directly involved in the running of the biobank might be possible and could be considered: a national or international oversight body might be able to monitor one or more biobanks. However, funding for such bodies might be problematic. As well, there is a question of which groups would be interested in taking on such a role.

Recommendations to the MSSS

In the resulting report to the MSSS (Bédard et al. 2007), the authors noted that how a biobank is governed largely depends on the elements of the common framework and any biobank-specific laws and guidance that exist. Québec already had in place the following governance mechanisms: the Canadian Constitutional Charter of Rights and Freedoms and the quasi-constitutional Québec Charter, the Québec Civil Code with dispositions on research as well as privacy laws (public and private) and the Canada Health Act 1985 with sections on privacy. There was also the Fonds de la recherche en santé recommendation on population biobanks (FRSQ 2006), the Canadian Tri-Council Policy Statement (TCPS 1998) and the Réseau de Médecine Génétique Appliquée statement of principles for human genetic research (RMGA 2000).

The authors recommended that if IPEG, or another body, were to act as a governing body for a biobank such as CARTaGENE, it should have the following structure:

1. Mandate – the body should have:

   a. An oversight role
   b. Input into the creation of policies and statements
   c. The ability to ensure that the biobank is following its protocols

2. Membership – members should include:

   a. Experts in ethics, law, epidemiology, genetics, sociology, informatics, security, database management, etc.
   b. One member representing the 'population'
   c. Two members representing interest groups, funding bodies, and the biobank
   d. International membership

3. Operations:

   a. Members should be named for a fixed period of time with a possibility of renewal
   b. Biobank representatives should be invited to meetings but not given the right to vote
   c. Body should meet at least four times per year

4. Finances:

    a. Funding should be public money from a governmental source.

After receiving this report, the MSSS decided not to appoint IPEG as an independent governing body for CARTaGENE. This may have been because the political context at that time was not conducive to creating the body. At that time, the MSSS was restructuring the research ethics evaluation process in Québec by creating a multi-center ethics review system. When fully in place, this new system is expected to dramatically transform the culture of ethics evaluation in the province of Québec. Thus, in the meantime, the MSSS decided that existing legislative and policy governance mechanisms, when combined, would provide sufficient oversight for biobank projects, such as CARTaGENE, in Québec. The IPEG was dissolved in 2008.

Conclusion

In the absence of IPEG, CARTaGENE staff have been following the rules and conditions set by the institutional research ethics committees, including the new multi-centric system for ethics review, and those of the Information Access Commission, a governmental body. The CARTaGENE governance structure has expanded; multiple committees have been formed to carry out specific roles, as well as liaise between the different project partners to ensure good communication. Le comité d'accès aux données et specimens (SDAC), a committee to govern requests for access to data and samples by outside researchers, has been created and Le comité d'éthique de la recherche de la Faculté de medicine de l'Université de Montréal (CERFM) is responsible for ethics review of the CARTaGENE project.

It will remain unknown whether CARTaGENE's existing governance structure provides better oversight and is more efficient than having a single governing body in IPEG. Biobanks and population biobanks, such as CARTaGENE, are relatively new entities, and it is clear that more research is needed into how and why they are governed as they are, and how those forms of governance have impacted their work. The diversity of governance structures that are in existence for population biobanks is a reflection of this.

Our analysis shows that while the biobanks studied shared a common governance framework (for example, scientific and ethical review; data protection laws; other domestic legislation; and ethical norms), there were significant differences that reflected the context in which the biobank was situated. No one governance framework could be determined that would fit all biobanks. With more time and experience, it may be possible to determine the optimal governance framework for a population biobank.

Bibliography

Bédard, K., Lazor, S., and Knoppers, B.M. 2007. Mécanismes de gouvernances pour les resources/biobanques populationelles: Recommandations pour le Québec.

Cambon-Thomsen, A. 2004. The social and ethical issues of post-genomic human biobanks. *Nature Reviews Genetics*, 5, 866-873.

Canada Health Act (R.S., 1985, c. C-6).

Canadian Charter of Rights and Freedoms, Canada Act 1982 (1982 c. 11) Schedule B (Constitution Act).

CARTaGENE 2009. *CARTaGENE Project FAQ*. [Online] Available at: http://www.cartagene.qc.ca/index.php?option=com_easyfaq&Itemid=49.

Civil Code of Québec (S.Q., 1991, c. 64).

Deschênes, M. and Sallée, C. 2005. Accountability in population biobanking: Comparative approaches. *Journal of Law, Medicine & Ethics*, Regulation of Biobanks (Spring 2005), 40-53.

EGC. 2009. *UK Biobank Ethics and Governance Council Frequently Asked Questions*. [Online] Available at: http://www.egcukbiobank.org.uk/faqs/index.html.

FRSQ. 2006. *Governance Framework for Data Banks and Biobanks Used for Health Research*. [Online] Available at: http://www.frsq.gouv.qc.ca/en/publications/autres_publications.shtml.

Generation Scotland. 2009. *Generation Scotland: Scientific Committee*. [Online] Available at: http://129.215.140.49/gs/gcommittee.htm.

Genome Canada. 2007. *Canada's New Government and Quebec Government Make one of the Largest Investments in International Genomics Research: Genome Canada, Génome Québec and Université de Montréal launch P³G Consortium and CARTaGENE Project*. [Online] Available at: http://www.genomecanada.ca/en/medias/news.aspx?i=279.

Gibbons, S.M., Helgason, H.H., Kaye, J., Nõmper, A. and Wendel, L. 2005. Lessons from European population genetic databases: Comparing the law in Estonia, Iceland, Sweden and the United Kingdom. *European Journal of Health Law*, 12(2), 103-33.

Greely, H.T. 2007. The uneasy ethical and legal underpinnings of large-scale genomic biobanks. *Annual Review of Genomics and Human Genetics*, 8, 343-364.

Institute on Governance. 2003. *Principles for Good Governance in the 21st Century*. [Online] Available at: http://www.iog.ca./publications/policybrief15.pdf.

JFHS. 2006. *Briefing Document: Privacy and Confidentiality Protections for Participants in the Joondalup Family Health Study*. [Online] Available at: http://www.jfhs.org.au/uploads/sk/zV/skzVmSBh_JlhyKCdxKy5YA/JFHS-Study-Protections.pdf.

Knoppers, B.M., Fortier, I., Legault, D., Burton, P.R. 2008. The public population project in genomics (P^3G): A proof of concept? *European Journal of Human Genetics*, 16, 664-665.

Nys, H. and Fobelets, G. 2008. The regulation of biobanks in Spain. *Law and Human Genome Review*, 29, 169-188.

Opinion Leader Research. 2003. *Summary of the UK Biobank Consultation on the Ethics & Governance Framework*. [Online] Available at: http://www.ukbiobank.ac.uk/docs/public-stakeholder-olr-report.doc.

Palmer, L.J. 2007. UK biobank: Bank on it. *Lancet*, 369(9578), 1980-2.

P^3G Observatory. 2008. *Lexicon*. [Online] Available at: http://www.p3gobservatory.org/lexicon/list.htm.

P^3G. 2009. *The Public Population Project in Genomics (P^3G)*. [Online] Available at: http://www.p3gconsortium.org/.

Quebec Charter of Rights and Freedoms (R.S.Q., c. C-12).

Seminara, D., Khoury, M.J., O'Brien, T.R. et al. 2007. The Emergence of Networks in Human Genome Epidemiology: Challenges and Opportunities. *Epidemiology*, 18(1), 1-8.

TCPS. 1998. Canadian Institutes of Health Research, Natural Sciences and Engineering Research Council of Canada, Social Sciences and Humanities Research Council of Canada 1998 (with 2000, 2002 and 2005 amendments). *Tri-Council Policy Statement: Ethical Conduct for Research Involving Humans*. [Online] Available at: http://www.pre.ethics.gc.ca/english/policystatement/policystatement.cfm.

Tutton, R., Kaye. J. and Hoeyer K. 2004. Governing UK biobank: The importance of ensuring public trust. *Trends in Biotechnology*, 22(6), 284-5.

Wallace, S., Bédard, K., Kent, A. and Knoppers, B.M. 2008. Governance mechanisms and population biobanks: Building a framework for trust. *GenEdit*, 6(2), 1-11. [Online] Available at: http://www.humgen.umontreal.ca/genedit.

Chapter 15
UK Biobank Ethics and Governance Council: An Exercise in Added Value

Martin Richards, Adrienne Hunt and Graeme Laurie

The purpose of UK Biobank is 'to establish and operate a resource for research with the aim of improving the prevention, diagnosis and treatment of illness and promoting health throughout society for public benefit' (UK Biobank 2007a). Recruitment of participants is currently underway and on completion in 2010 the resource is expected to contain health and lifestyle data and biological samples from some 500,000 people from the UK who were aged 40-69 at the time of recruitment. During the planning of the project, and in parallel with the development of the scientific protocol, an Interim Advisory Group (IAG) was established to make recommendations to the funders about an ethics and governance framework for the project. Two key recommendations to emerge from the deliberations of the IAG were, first, that UK Biobank should adopt an Ethics and Governance Framework (EGF) which lays out explicitly the commitments of UK Biobank to its participants, the public and other stakeholders, and, second, that a permanent and independent Ethics and Governance Council should be established to oversee the project and to monitor and advise on its operation. As a result, a permanent Ethics and Governance Council (EGC) was formed in 2004. In this chapter we will discuss the reasons why it was thought necessary to create this body, in addition to the other institutions and processes that existed for the regulation and ethics governance of research. We will comment on the practice and work of the EGC in light of our experiences of working on the Council and talk about the ways that the EGC has dealt with specific issues. But first we will provide a little more detail about the UK Biobank itself.

UK Biobank[1]

UK Biobank is a charitable company limited by guarantee and is the legal owner and steward of the database and sample collection. The broad purpose of the project is to establish a resource that can be used by many different researchers for medical and health-related research. Once recruitment is complete (expected in 2010) the resource will be available to medical and health researchers in both

1 See www.ukbiobank.ac.uk for further information.

the public and private sectors, subject to successful application and appropriate approval processes. The rȯle of UK Biobank therefore, is not to carry out research but to be responsible for creating and maintaining the resource, and to regulate the type of access that is given to researchers.

Recruitment to UK Biobank is being organized via a series of assessment centres, each operating for about six months in a chosen location. Those in the locality of the assessment centre and in the relevant age group are invited to join the project by letter. The letter includes an information leaflet and a provisional appointment which can be accepted, changed for a more convenient time, or simply ignored. Initial contact information about potential participants is obtained through accessing the contact details from National Health Service Records of people who live in a specific area. This process has received ethics approval and legal approval from the Patient Information Advisory Board (PIAG) and involves only the minimum amount of personal details needed to contact eligible persons. According to the EGF: 'UK Biobank will seek to recruit as widely generalizable a population sample as is practicable so that the research may ultimately benefit a wide diversity of people' (UK Biobank 2007b: Section 1.A.2). It has been reported by UK Biobank that the proportion of those written to who are recruited to the project varies in general between 6-12 per cent between assessment centres but is about 9 per cent overall (UK Biobank EGC 2008a). A high attendance rate of 20 per cent has been reported for Bristol (UK Biobank EGC 2008b). At the time of writing (March 2009) over 270,000 people have been recruited.

At each assessment centre would-be participants are given further details of what involvement in the project entails. It is made clear that consent is being given 'to participate in UK Biobank' with its broad purpose to create a resource for health-related research. At this stage, individuals cannot be informed of all the research uses, or the names of researchers and research groups who may seek access to the resource. Therefore, participants are asked to give a broad consent to different research uses, having been informed about the purpose and nature of UK Biobank and the way that it will operate and be governed. As a part of this, they are given an undertaking that only anonymized data or samples will be provided to researchers for research purposes.

During the 90 minutes or so of the assessment, participants provide information about health, lifestyle, memory, work and family history through a touch screen questionnaire and health interview; undergo some physical measurements (including blood pressure, pulse rate, height, weight and bone density); and provide biological samples (blood and urine) (see Sample and Tutton 2008, UK Biobank 2006 for accounts of experience of recruitment). Participants receive some details of their own physical measurements but are told that thereafter there will be no further feedback of any individual information, such as the individual results of research conducted on the resource. At some of the assessment centres there is also a nurse provided by the British Heart Foundation with whom participants can discuss their cardiac assessments. The British Heart Foundation chose some of the larger centres where they expected more participants from low socioeconomic backgrounds to test whether their screening programme might be able to engage people at high risk of cardiovascular disease who would not usually have a risk assessment.

Further data on each participant will be collected from their NHS records and other health-related records (for example the National Cancer Registry) and there may be future re-contact to make further assessments from some or all of those enrolled. From the outset, UK Biobank has been envisaged as a long-term project which will continue to collect information about participants and which is expected to hold data and samples over many decades. Information about the progress of the project and results of research carried out using the resource will be made available to participants and society in general, largely via the UK Biobank website and through the publication of research findings.

Research Governance and Regulation

As with any research project involving human participants, UK Biobank is subject to the UK national framework of research governance and regulation. It falls within national legislation, for instance, the *Human Tissue Act 2004* and *Data Protection Act 1998*, common law duties of care and confidentiality, and is also under the watchful eyes of a number of regulators, such as the Human Tissue Authority, the Information Commissioner's Office, the Charity Commission and Companies House. Other relevant governance mechanisms include the UK National Health Service, which acts as a data controller with respect to information about participants; the system of Research Ethics Committees which will review relevant research projects involving UK Biobank; as well as the organization of the UK Biobank itself. Professional ethics and guidelines (for example, the Medical Research Council) are also involved and, last, but not least, there is the freely-given consent of all participants.

It was decided early in the planning of UK Biobank that an ethical framework for the project should be created in parallel with the development of the scientific protocol, and in addition to the pre-existing governmental regulations to which the project would also be subject. This was because of the uniqueness of the proposal in terms of the size (both the number of participants and the amount of data and biological material, including DNA, to be held on each) and the complexity of the project. It was felt that an ethical framework tailored specifically for UK Biobank would foster and maintain public trust in the project. A widespread consultation with health and science professionals, experts in ethics and law and members of the public was carried out to solicit ideas on what the parameters of the ethical framework should be. As a result an Interim Advisory Group on Ethics and

Governance was set up by the funders.[2] This activity resulted in the development, publication of, and consultation on an Ethics and Governance Framework.

The EGF (UK Biobank 2007b) provides an ethical framework for UK Biobank and the establishment, use and management of the resource. This document provides a detailed, bespoke approach to the ethical issues that emerge from running a biobank of this size within the UK. It establishes that consent will be sought to participate in UK Biobank. This will be based on an explanation and understanding of a number of features of participation (such as the kinds of information and samples that will be collected at enrolment and the possibility of being re-contacted in the future by UK Biobank). There is also a general description of the potential research uses of the data and samples, but broadly UK Biobank will encourage use of the resource for health-related research. The EGF affirms the right of participants to withdraw at any time and makes a commitment to protecting the confidentiality of both samples and data. UK Biobank is confirmed as steward of the resource and legal owner of the database and sample collection. The principles which govern access to the resource by researchers are described. The framework goes on to outline the benefit-sharing that will be required which includes the obligatory publication of findings and accessible archiving of research findings in UK Biobank for future use.

Finally, it describes the arrangements for management and accountability. These include the creation of an Ethics and Governance Council (EGC). The EGC was set up in 2004 by the principal funders of UK Biobank, the Medical Research Council and the Wellcome Trust. Its role was to act as an independent guardian of the UK Biobank Ethics and Governance Framework (EGF), to advise on its revision, to monitor and report publicly on conformity of the project with the EGF and, more generally, on the interests of research participants and the general public in relation to UK Biobank.

The Rationale behind the Ethics and Governance Council

We might begin by asking why the funders set up an independent EGC; why not simply rely on the ethics governance processes that cover all UK research? The consultations that were carried out in the process of formulating the EGF and on the draft document itself strongly supported the setting up of an independent body. It was a particular recommendation from the Interim Advisory Group that there should be a body to monitor UK Biobank's conformance with the EGF.[3] It was

2 The membership of the Interim Advisory Group was: Dr William Lowrance (Chair), Professor Alastair V. Campbell, Professor Erica Haimes, Dr Graeme Laurie, Professor Chris Mathew, Professor Jean McHale, Mrs Helen Millar, The Baroness O'Neill of Bengarve, and Mrs Madeleine Wang.

3 Indeed, during the early development of UK Biobank a Government White Paper envisaged the Council not only as an advisory body but as a committee that has 'the

generally felt that the existing medical research governance framework was not sufficiently able to deal with some of the unique issues raised by such a project. These concerns were largely to do with the fact that UK Biobank was going to be established as a resource, which did not fit neatly within the medical research governance framework that had been developed for single purpose, hypothesis-led projects.

There are a number of features of UK Biobank which led to the recommendation to establish the EGC. UK Biobank has adopted a broad model of consent from participants. As with all such biobanks the future uses of the resource cannot be foreseen at the point at which a participant is recruited. So, the key purpose of the EGC is to monitor that UK Biobank is always managed in conformity with the original consent of its participants and to keep under review applications for access to the resources with regards to the interests of UK Biobank participants. In advising on the participants' interests in the future, for example, the Council might ask itself if a 'reasonable participant' would have expected the proposed use of his or her data or samples at the point at which they gave their consent. We might note in parentheses that a number of current members of the EGC have happened to live in locations where UK Biobank has been recruiting and have accepted the invitation to take part in the project, which enables them to have a clearer understanding of the issues that are faced by participants in the project.[4]

Another factor that led to this decision is that UK Biobank is a long-term project which will evolve over time. This raises challenges for existing regulatory mechanisms because traditional research ethics committee (REC) approval – required of all research involving human subjects at the beginning of a project – is a one-off event. Active and on-going monitoring of a project's progress is not generally within a REC's brief as they are not organized in a manner that would allow them to take on this role and neither do they have the resources to carry out such monitoring on a regular basis. This is where a body such as the EGC can add value to a governance model by actively monitoring and developing alongside the scientific project.

Of course, UK Biobank is not unique for being a long-term project. For example, Britain has a series of birth cohort follow-up studies which recruited a week's births in 1948, 1956 and 1970 which continue to follow these people – and indeed their descendants today. These, like UK Biobank, have also relied on broad consent (of the mothers, in the first instance). However, they have only recently collected DNA from participants. This is a feature of UK Biobank that has led to concerns by some commentators about the use of broad consent and the protection of privacy (see Caulfield 2004, Kaye 2004). However, despite these similarities

---

power to veto uses of the data or samples that it considers to be against the interests of the participants or likely to damage the reputation of the study' (Department of Health 2003).

4 During the consultation there was wide support for the idea that participants should be represented on the EGC though membership does not expressly include participant representatives. The EGC was initially set up before recruitment began.

with UK Biobank none of these cohort studies has its own independent ethics and governance body. An additional feature of UK Biobank that makes it different to the cohort studies is that it is considerably larger in terms of participant numbers, as it will be 10 times the number of participants of the other birth cohort studies.

Moreover, UK Biobank was set up at a time when social attitudes towards medical research and concerns about the protection of personal interests were different to those which existed previously. Notably, public attitudes about what it means to be a research participant and which principles and parameters should govern issues such as data protection and access have led to a changing social context – and thus, it might be argued, have necessitated a governance structure that can reflect and respond to such change. An important reason for the establishment of an independent EGC was to foster and maintain public trust in the UK Biobank project through its advisory and monitoring roles. A report from one consultation, 'Biobank UK: A Question of Trust: consultations exploring and addressing questions of public trust' (People, Science and Policy Ltd. 2002) recommended that 'some form of oversight body should be established and that the body should be capable of acting independently of the user and sponsors'. Any such body 'should ensure that standards of behaviour and ethics are maintained and continued to reflect the public mood as the public consensus changes but within the original terms of consent given by volunteers' (People, Science and Policy Ltd. 2002: 4).

A later consultation on the draft EGF was more specific. An oversight body was seen to be critically important and should be composed of 'professional people, who knew what is going on' (People, Science and Policy Ltd. 2003) as well as lay and participant representation with recruitment following open and transparent procedures. As we have indicated, the Ethics and Governance Council was established in November 2004.

In Practice: The EGC and its Work

Members of the EGC are appointed through processes in keeping with the Nolan Principles on Standards in Public Life, which requires that there should be a public invitation of applications and members should be appointed on the basis of their backgrounds and abilities to further the purposes of the Council. The currently-constituted Council is a multi-disciplinary group including expertise in law, ethics, biomedical sciences, policy, social science and consumer issues.[5] It has 10 members and meets four times a year. In addition, it has one or more public meetings per year. Recently, these have been in localities where UK Biobank

5 Details about membership of the Council, minutes of its meetings and other information can be found on its website, available at: http://www.egcukbiobank.org.uk. It is independent from UK Biobank though is funded by the UK Biobank funders who also approve appointment of members.

is recruiting. These meetings have attracted very mixed groups including some who have been recruited to UK Biobank, as well as others who had declined the invitation to take part.

The important feature of the EGC is that it is an independent body which 'speaks about UK Biobank, not for UK Biobank'. It does not assume responsibility for the ethical management of the resource – that remains the responsibility of UK Biobank itself. Nor does it own or develop the EGF and associated policies, it can only advise on their formulation.

When advising, reviewing and reporting on UK Biobank's activities, the EGC serves as a mirror for UK Biobank, providing critical and constructive advice. As stated in the Ethics and Governance Framework 'normally the Council will communicate its reflections and criticisms informally. If the Council is not satisfied with UK Biobank's response, it could make a formal statement of concern to the Board or funders, or, if necessary, make a public statement that certain actions should or should not be taken' (UK Biobank 2007a: Section III.A.2). However, it has no formal mechanisms, such as fines or enforcement orders for ensuring that its advice is followed.

In order to maintain this role, an important feature of the relationship between UK Biobank and the EGC is good communication and the free flow of information. At each meeting, the Principal Investigator of UK Biobank (who is currently Professor Rory Collins of the University of Oxford) generally attends for part of each Council meeting and reports on the current progress of UK Biobank. Papers on particular issues or future plans may be provided by UK Biobank or requested by the EGC. Other UK Biobank officers may attend to describe and discuss their aspects of the project. For example, the officer responsible for information technology security provided a detailed description of the architecture of the data storage system and its security, and how it has met tests to assess its robustness in the face of trial attempts to gain access to information. The Council has also seen aspects of the work first hand, visiting an assessment centre and the central sample processing and storage facilities.

The Changing Nature of the EGC's Work

The EGC was established at a time when UK Biobank was developing its protocol and establishing its procedures. Today the project is recruiting approximately 600 people per day, six days a week at six dedicated assessment centres throughout the UK. During this period the EGC has been both reactive and proactive to the project's stage of development. As such there has been a change in the focus of the EGC's work over the last three years as different aspects of its remit come to the fore. Initially, the Council engaged predominantly with its advisory function. This included advising on revisions to the EGF itself and to the policies and procedures that flow from the framework. For example, the Council has advised on the standard operating procedures for the management of incidental findings at the assessment centres and for judging a potential participant's capacity to consent. The Council also reviewed and advised extensively on the project's participant information leaflet and consent form in order to assure itself that the elements of consent described in the EGF were conveyed in these materials. This involved informing participants explicitly that the range of potential researchers who might use the resource included researchers from other countries and those with commercial links. During this period the Council also advised on the standard operating procedures for the handling of complaints and enquiries received by the project. In reviewing and advising on these materials the Council wishes to assure itself that UK Biobank has effective mechanisms in place to deal with participants' complaints and enquiries in a prompt and efficient manner.

Over the last year (2007-2008), the Council has started to engage more with the monitoring aspect of its remit. For example, it has moved from providing advice regarding the standard operating procedures for complaints and enquiries handling, to monitoring the actual complaints and enquiries received by the project. These are monitored through biannual reports prepared by UK Biobank. The reports outline the number and types of enquiry and complaint and also detail the reasons given for participant withdrawals. The Council is interested in monitoring both the reason for, and trends in, the complaints and enquiries received by the project. These could provide an indication of how happy, or unhappy, individuals are with certain aspects of UK Biobank – whether that be the practical aspect of the assessment centre operation (such as waiting times or people having difficulty with some of the questions in the questionnaire) or about certain policies that are found to be unacceptable, for example the feedback policy. In the longer term, the Council will look at the trends in the complaints and enquiries to see if certain concerns or opinions shift with time – possibly indicating a shift in concerns and opinions in society more broadly. This could cover the possible future uses of the resource if certain types of research become more or less publicly acceptable over time (see also Laurie, Bruce and Lyall 2009).

Commissioning Research

The last function we would like to mention, in addition to the advisory and monitoring role, is a foresight role. In the early days of the Council its role was more reactive, reviewing and advising on procedures and policies that UK Biobank was developing. Over the last year the Council has started to be more proactive about identifying and investigating issues which it thinks warrants further consideration either by UK Biobank or by the Council itself. A notable example is the challenges yet to come with respect to access to the resource and the management of any intellectual property that is generated from that access. While access is not anticipated before recruitment completes in 2010, the Council has none the less established a sub-group to anticipate and examine issues likely to arise. As part of this foresight role the EGC has commissioned research in order to help it identify and investigate emerging issues.

In 2006 the Council commissioned a scoping study of public attitude surveys regarding biobanking related issues. This scoping study (Sumner 2007) identified a gap in the literature regarding public attitudes to certain aspects of access including opinions regarding commercialization and benefit-sharing. The Council has since commissioned a further study to investigate actual public attitudes with respect to these areas generally and in the context of current UK Biobank policy. This study will help to test the Council's own understanding of these issues and will in turn help the Council in formulating its advice to UK Biobank regarding the further development of its access procedures (Webster et al. 2008).

In addition to research regarding public attitudes, the Council has commissioned a study to provide a conceptual analysis of the terms the 'public interest' and the 'public good'. These terms are often used in the context of UK Biobank, including the EGF which describes UK Biobank as being maintained and built 'for the public good' (UK Biobank 2007a). However, as a Council we felt that the application of these very general terms to the management of resources like UK Biobank was not sufficiently clear or well articulated in order to be useful for policy making.

In commissioning this research the EGC wanted to inform itself of how these phrases might be interpreted in the context of UK Biobank and how they might apply to decision-making about who should have access to the resource and on what basis. Ultimately the EGC would like to develop a framework of principles to guide its future advice and decision-making with respect to access requests to the UK Biobank resource. The conceptual analysis was commissioned in order to inform, and provide a theoretical base for, the development of this framework.

This piece of work was reported to us in 2007 and is available on the EGC website (Capps et al. 2008). In order to extract the most value from the report, and in keeping with its commitment to transparency of advice and decision-making, the EGC has produced a document which summarizes the findings of the report in the context of their application to the EGC's activities (UK Biobank EGC 2008). The summary is a 'living' document and will be revised over time to take account of new thinking by the Council on its interpretations of 'public interest' and 'public good' (e.g. in response to other commissioned work or future experiences).

Providing Advice

Finally, there are two additional areas where the Council has, and is, in the process of advising on issues that were either not foreseen or areas that are actively being developed. We believe these further demonstrate how a body like the EGC adds value and complements existing governance and regulatory mechanisms.

The first example relates to the method by which a participant can withdraw from the project. Three levels of withdrawal are offered (UK Biobank 2007b). These range from 'No further contact' which means that UK Biobank would no longer contact the participant directly, but would still have their permission to use information and samples provided previously and to obtain further information from their health-relevant records; 'No further access' which means that UK Biobank would no longer contact the participant or obtain further information from their health-relevant records in the future, but would still have their permission to use the information and samples provided previously and 'No Further Use' which was described in previous drafts of the EGF and information materials as follows:

> In addition to no longer contacting the participant or obtaining further information, UK Biobank *will destroy all of their health-related information* and samples collected previously (although the participant would be told that it may not be possible to trace and destroy all distributed anonymised sample remnants). Only some administrative details (such as their signed consent and withdrawal) would be kept as a record of their wishes. Such a withdrawal would prevent information about them from contributing to further analyses, but it would not be feasible to remove their data from analyses that had already been done. (emphasis added)

In June 2007 Professor Collins, the Principal Investigator for UK Biobank, sought the Council's advice regarding this option of withdrawal because it had come to light that it is not possible for UK Biobank to destroy all of a participant's health-related information due to the project's data back-up and audit system. Importantly, however, UK Biobank can still guarantee the main principle behind this option, i.e. that there will be no further use of the data by researchers.

Following initial discussions the Council requested and received a briefing note from UK Biobank describing in detail the technical problem surrounding the withdrawal of data from the resource. Also, UK Biobank's Systems Architect attended a Council meeting to provide an update on the project's Information Technology and Data Management Strategy and to describe the technical and physical procedures involved when a participant withdraws from the project.

After further discussion the Council concluded that it considered that the fundamental guarantee described in the EGF is that there will be no further use of the data by researchers. This guarantee has not changed. The Council subsequently made several recommendations to UK Biobank that resulted in the following actions:

- The EGF and participant information leaflet were revised to describe the 'No further use' withdrawal option as follows:

> This means that, in addition to no longer contacting you or obtaining further information about you, *any information and samples collected previously would no longer be available to researchers*. UK Biobank would destroy your samples (although it may not be possible to trace all distributed sample remnants) and *would only hold your information for archival audit purposes*. Your signed

> consent and withdrawal would be kept as a record of your wishes. Such a withdrawal would prevent information about you from contributing to further analyses, but it would not be possible to remove your data from analyses that had already been done. (emphasis added)

- Additional text was inserted into the EGF describing UK Biobank's commitment to keeping participants informed about important developments with the project.
- A new page was added to UK Biobank and the EGC's website and information about this change was posted on both sites. Further, these pages will be updated with any such policy or procedure changes in the future.

This example illustrates the likelihood that issues will arise that have not been anticipated but that require ethical consideration. In this example the Council tried to consider the situation in terms of the participants' expectations and in so doing aimed to provide UK Biobank with an appropriate and proportionate response that is acceptable to participants and the public.

Our final example relates to an area of ongoing discussion between UK Biobank and the EGC: the policy on providing participants with feedback of health information resulting from their participation in the project. During the developmental stages of the project the Council reviewed and advised UK Biobank on its policy and standard operating procedures in relation to this topic. Currently feedback is provided for a number of measures taken at the assessment centre but no feedback is provided on information that is derived about a participant after the assessment centre visit. So, at present, participants only rcccivc information about smoking, blood pressure, body size, heel bone density and lung function. These individual measures are reported with reference to population standards (thus avoiding individual interpretation and aiming, therefore, to stay true to the fact that UK Biobank is a research project, rather than a clinical encounter). There is also a standard operating procedure on how to manage incidental findings made during the baseline assessment. Under this, if staff notice something that could be significant (for example, abnormal skin colouration observed on the arm during blood collection which could be indicative of a potential melanoma) staff enquire if the participant is aware of it and, if not, it is suggested they ought to make an appointment to see their GP.

When UK Biobank's funding was being considered an international scientific advisory board was convened by the funders to peer review the project's protocol. The board endorsed UK Biobank's protocol and in addition made a series of recommendations. One recommendation was that UK Biobank could increase the value of the resource by collecting additional data on a subset of the cohort.

In response to this recommendation, UK Biobank convened a sub group to look at the types of measures and questions that could be added to enhance the baseline measures and questions that are already administered to all participants.

The conclusions of this sub-group informed a proposal that has recently been put forward to the funders and the EGC for initial consideration.

The new measures range from additional questions and measures regarding exercise, diet and exposures to new measures regarding hearing and sight through to magnetic resonance imaging of the body. We will not go into a great deal of detail about these new measures because they are only at the proposal stage. However, the proposals have raised certain issues which UK Biobank and the Council will be considering in the coming months. In particular, the implications of the proposals in relation to UK Biobank's current policy on providing – or not providing – feedback of health information to its participants will be considered.

This example shows that certain policies cannot be taken for granted. That is, it cannot be assumed that a current policy will necessarily be appropriate as a study develops. The EGC intends to go back and review its original advice to UK Biobank and to re-assess the principles that underlie the current policy on feedback. In due course we hope to provide consistent, informed advice back to UK Biobank about the further development of its protocol.

Conclusions

In conclusion, the Council was established in order to build and maintain public trust and in direct response to the broad model of consent adopted by UK Biobank and to the long term nature of the resource. The Ethics and Governance Framework (EGF) and the Council are intended to provide an extra safeguard for participants and the public and both are intended to provide a foundation for trust.

The Council aims to build this trust through its advisory and monitoring role and through the public reporting of its own activities and of UK Biobank's conformance with the Ethics and Governance Framework. The monitoring aspect of the Council's remit is of particular importance as a complement to traditional research ethics committees which do not typically offer a complete monitoring role. Through its work the Council hopes to advise on the public interest in relation to the project as it may change over time and to assure itself that UK Biobank is operating in the terms of the EGF and the participants' consent.

As projects like UK Biobank develop there will be changes – both foreseen and unforeseen – that will require consideration. A body like the Council is able to provide a consistent, informed view on these situations, acting in the public interest and keeping key principles at the heart of its decisions. Through its work a body like the Council can promote and facilitate good governance and good science in the public interest.

Bibliography

Capps, B. et al. 2008. *Access to the UK Biobank Resource: Concepts of the Public Interest and the Public Good. London: UK Biobank Ethics and Governance Council.* [Online: EGC UK Biobank] Available at: http:www.egcbiobank.org.uk/meetingsandreports [accessed 22 September 2008].

Caulfield, T. 2004. Perceptions of risk and human genetic databases: Consent and confidentiality policies, in *Blood and Data: Ethics and Social Aspects of Human Genetic Databases*, edited by G. Arnason et al. Reykjavik: University of Iceland Press, 283-289.

Department of Health. 2003. *Our Inheritance, Our Future. Realising the Potential of Genetics in the NHS.* [Online: Department of Health] Available at: http:www.dh.gov.uk/en/Publicationsandstatistics/Publications/PublicationsPolicyAndGuidance/DH4006538 [accessed 22 September 2008].

Kaye, J. 2004. Abandoning informed consent: The case of genetic research in population collections, in *Genetic Databases: Socio Ethical Issues in the Collection and Use of DNA*, edited by R. Tutton and O. Corrigan. London: Routledge, 117-138.

Laurie, G., Bruce, A. and Lyall, C. 2009. The roles of values and interests in the governance of the life sciences: Learning lessons from the 'ethics+' approach of UK Biobank, in *The Limits to Governance: The Challenge of Policy-making for the New Life Sciences*, C. Lyall et al. (eds) Aldershot: Ashgate, 51-77.

People, Science and Policy Ltd. 2002. *UK Biobank: A Question of Trust: A Consultation Exploring and Addressing Questions of Public Trust.* [Online: UK Biobank] Available at: www.ukbiobank.ac.uk/ethics/consult.php [accessed 22 September 2008].

People, Science and Policy Ltd. 2003. *UK Biobank: Consultative Ethics and Governance Framework.* [Online: UK Biobank] Available at: www.ukbiobank.ac.uk/ethics/consult.php [accessed 22 September 2008].

Sample, M. and Tutton, R. 2008. Biobank as biographical disruption: Conversations on some first person reflections. *Medical Sociology*, 3(2), 15-24 [Online].

Sumner, J. 2007. *Public Attitudes to Biobank and Related Ethics and Governance issues.* London: UK Biobank Ethics and Governance Council. [Online: EGC UK Biobank] Available at: http://www.egcbiobank.org.uk/meetingsandreports [accessed 22 September 2008].

UK Biobank. 2006. *Integrated Pilot Report.* [Online] Available at: http://www.ukbiobank.ac.uk/ethics/intro.php [accessed 22 September 2008].

UK Biobank. 2007a. *UK Biobank Ltd. Annual Review (2007).* [Online] Available at: http://www.ukbiobank.ac.uk/about/what.php [accessed 22 September 2008].

UK Biobank. 2007b. *UK Biobank Ethics and Governance Framework Version 3.* [Online] Available at: http://www.ukbiobank.ac.uk/ethics/intro.php [accessed 22 September 2008].

UK Biobank Ethics and Governance Council. 2008a. *Fifteenth Meeting Report.* [Online]. Available at: http://www.egcukbiobank.org.uk/meetingsandreports [accessed 22 September 2008].

UK Biobank Ethics and Governance Council. 2008b. *Sixteenth Meeting Report.* [Online] Available at: http://www,egcukbiobank.org.uk/meetingsandreports [accessed 8 March 2009].

UK Biobank EGC. 2008. *Access to the UK Biobank Resource: Advisory on the Public Interest and the Public Good.* [Online] Available at: http://www.egcbiobank.org.uk/meetingsandreports [accessed 22 September 2008].

Webster, A. et al. 2008. *Public Attitudes to Third Party Access and Benefit Sharing: Their Application to UK Biobank.* Biobank. London: UK Biobank Ethics and Governance Council. [Online] Available at: htpp://www.egcbiobank.org.uk/meetingsandreports [accessed 22 September 2008].

Chapter 16

The End of Individual Control Over Health Information: Promoting Fair Information Practices and the Governance of Biobank Research

Trudo Lemmens and Lisa Austin

Until recently, rules about protection of health information and about the protection of human research subjects largely existed in different universes. In the last couple of years, however, we can see a growing awareness of privacy concerns in the context of health research. This is associated with the overall growth of health information based research, but also with the increase in new forms of research such as genetic research. With health information and genetic research we are not dealing with the risk of potential physical side effects of novel therapies but with informational risks. Gene sequencing studies or research involving genetic biobanks raise particular concerns about the misuse of information by third parties, and about the potential psychological harm of information.

Research ethics guidelines and regulations have started including more detailed provisions on privacy protection. But they have largely continued to emphasize the importance of informed consent as the main mechanism in the context of sharing health information. At the same time, some jurisdictions have introduced health information privacy legislation to address in part the growing informational privacy risks in research. Some Canadian provinces have incorporated traditional research ethics mechanisms of independent review by Research Ethics Boards (REB)[1] in health information privacy legislation (e.g. *Personal Health Information Protection Act* 2004 [Ont.], *Health Information Act* 2000 [AB.]). These statutes are modelled upon 'Fair Information Practices' but nonetheless allocate a specific role to REBs, which have been developed in the context of another regulatory regime.

In this chapter, we want to discuss how some Canadian provinces have promoted the integration of the concept of Fair Information Practices into the existing system of research ethics review, without appropriate recognition of the structural weaknesses of the current regulatory regime surrounding research. This

---

1 Research Ethics Boards is the Canadian term for what other countries term Research Ethics Committees or Institutional Review Boards. Since we discuss Canadian privacy statutes in some detail, we will use the term REB.

leads to the following somewhat contradictory situation: privacy statutes have joined the complex apparatus designed to protect research subjects and the public. They have introduced the concept of Fair Information Practices, well-established in privacy law, in the context of medical research, which diminishes the role of traditional informed consent procedures as the sole protective mechanism. Yet, privacy statutes presume at the same time that there are adequate governance systems in the context of research ethics review that can ensure the elaboration and enforcement of Fair Information Practices. This development is particularly important in the context of biobank governance. Indeed, health information privacy legislation will govern many of the aspects of the collection, use and disclosure of information associated with these biobanks. These privacy statutes allow for some leniency with respect to the informed consent requirements that could be difficult to satisfy in the context of biobank research. Yet, they do not provide a detailed structure for the committees that have to promote and enforce these Fair Information Practices. They contain very general provisions on REB review, which are insufficient as a basis for a solid regulatory structure surrounding research. We want to look whether from a perspective of 'good governance' principles, this system of REB review is currently appropriately structured in Canada to carry out the important function required under privacy legislation. Finally, we will discuss a recent proposal put forward in British Columbia, which entails the development of a specialized health information review committee. We want to contrast this model with the model of relying on non-specialized REBs that exists in many of the other provinces. The analysis is relevant for other jurisdictions around the world that have failed to introduce strict legislative or regulatory regimes for biobanks and instead may rely on inadequately regulated research ethics review systems.

The Limits of Informed Consent Associated with the Nature of Biobank Research

The clash between legal and ethical standards of informed consent and biobank practice

The development of biobanks confirms that there are solid reasons for focusing on governance systems in the context of health information privacy, and not to rely solely on informed consent as a protective mechanism. The emphasis on governance is in the first place appropriate because of the fundamental challenges of informed consent and the specific nature and characteristics of biobank research. As has been well documented and discussed elsewhere, legal and ethical informed consent requirements are difficult if not impossible to respect in biobank research (e.g. Kaye 2004, Caulfield 2002, 2007, Deschênes et al. 2001, Godard et al. 2003, Clayton et al. 1995, Greely 2001, Caulfield, Upshur and Daar 2003, Luther and Lemmens 2007). In addition, genetic research in general, and biobank research in particular, is associated with various ethical, legal, and social issues that are not easily addressed through traditional informed consent procedures. Informed consent faces practical, as well as conceptual problems in the context of biobanks. To understand the practical problem, one only has to contrast the nature and practices in biobank research with the requirements of informed consent.

The Canadian Longitudinal Study on Aging and CARTaGENE, to mention two Canadian examples, are population biobank projects that aim to follow people over their lifetime, to study various gene-gene and gene-environment interactions (e.g., Luther and Lemmens 2007). As with other biobank projects, they are clearly longitudinal and fundamentally open-ended and they continue to use a person's health information and biological samples after they have died. They involve the collection of a wide variety of data: clinical, biological, genetic, environmental, social, and other data. They cannot specify in advance for each sample what form of research will be undertaken in the future, what type of data will be used and connected to each other, and what researchers will be involved in the future. It is furthermore not possible to identify very precisely the type of informational risks that can result from research. Finally, the informed consent requirement in research usually also entails a duty of researchers to respect the right to withdraw. It is not as easy to withdraw consent for the use of a sample and of health information if these have been used for a long period of time and in very different settings. It is worth pointing out that some of the projects also insist that donors will not be able to remove health information and samples after withdrawal from the project.[2]

As in other jurisdictions, the traditional informed consent requirements of Canadian law and research ethics policies seem hard to reconcile with biobank practices. The research ethics policy document of the federal funding agencies, the Canadian Tri-Council Policy Statement (TCPS 2008), specifies that research subjects should receive the following information: a comprehensible statement of the research purpose, the identity of the researcher, the expected duration and nature of participation and a description of the research procedures (Art. 2.4.b); a comprehensible description of the reasonably foreseeable harms and benefits (Art. 2.4.c); an assurance that they are free not to participate, have the right to withdraw at any time, and will be given continuing and meaningful opportunities for deciding whether or not to participate (Art. 2.4.d); and the possibility of commercialization of research findings (Art. 2.4.e). While some general information on some of these points can be given (for example, on commercialization opportunities) it seems clearly difficult to provide much detail in the context of long-term biobank research.

The TCPS contains also provisions adapted to the reality of genetic research, banking of genetic material, and human tissue research. These provisions are,

2 See, for example, the information on consent and withdrawal on the website of the Ontario Health Study, a long-term chronic disease research project. Available at: http://www.ontariohealthstudy.ca/client/ohs/ohs.nsf/web/consent+and+withdrawal. 'We will continue to analyze any data or samples you have already provided. We will also continue to follow your health using pre-existing, administrative databases, such as the Ontario Cancer Registry'.

however, either very general, or too specific and perhaps not adapted to the increased possibility of de-anonymization, which will be briefly discussed further. Researchers who want to conduct research involving banked genetic material have to show that they 'addressed the associated ethical issues, including confidentiality, privacy, storage, use of the data and results, withdrawal by the subject, and future contact of subjects, families and groups' (Art. 8.6). It is also suggested in the discussion of this provision that the storage term should be defined. The discussion some agreement amongst researchers that such terms could be as short as five years or as long as 25 but these time limits seem hard to reconcile with actual biobank practices (Art. 8.6).

The TCPS provisions dealing with human tissue emphasize the need for 'free and informed consent' (and thus the principles laid out in the provisions on informed consent discussed earlier) (Art. 10.1(b)). Further details are provided as to the minimum information that donors of tissue must receive. Some of the required details are clearly difficult to provide in the context of biobank storage. For example, researchers have to give information on: the purpose of research; the location where tissue is to be taken; the potential uses for the tissue; identifying information attached to the tissue; and how the use of the tissue could affect privacy (Art. 10.2).

The common law requirements of informed consent in Canada are also in tension with biobank practices. The common law imposes a duty to disclose all information a reasonable person in a similar position would require to make an informed decision. In the context of research, this appears to require at a minimum details about the nature of the procedures, the specific risks, and the ability to withdraw consent. A few existing Canadian precedents further suggest that a higher duty to inform exists in the context of research, and that all risks, however remote they may be, have to be disclosed (*Halushka v. University of Saskatchewan* [1965] 52 WWR 608, *Weiss v. Solomon* [1989] A.Q. No. 312, Hadskis 2007).[3] The reason for this higher duty of disclosure resides in the fact that people in a therapeutic context can be presumed to accept some level of risk in order to receive treatment, and that they will also be generally familiar with the standard risks associated with therapy. The same is not true in the research context.

Several authors have argued for alternative informed consent models adapted to the reality of biobanking. One interesting approach is to use an option model, which gives donors of genetic material more control over the use of their material by providing them with a set of core choices, thus sketching in more detail for researchers the parameters surrounding the use of samples (Deschênes et al. 2001, National Bioethics Advisory Commission 1999, McGuire and Gibbs 2006). The proposed options include refusing some forms of research; allowing only

3 Hadskis makes the appropriate point that this higher standard of informed consent in research can be seen as an application of the general informed consent requirement, that is, a reflection of the idea that a reasonable person simply expects to receive more information in the context of research.

unidentified use; allowing the use of a coded sample with an option to contact research subjects for future use; allowing research without the option to re-contact; and a general authorization for future research. Other authors have recommended the use of an Authorization Model, again with recommendations for providing some options for future use. The use of the language of authorization reflects a realization that true consent is largely absent and that people can only be asked to give a general 'authorization' of future use of their samples (Caulfield, Upshur and Daar 2003). Finally, two Canadian privacy experts propose broadening the definition of health care to include some types of research that have direct clinical applications and then using a model of implied consent (Kosseim and Brady 2008). This idea could also be connected to a larger 'social contract' idea associated with being a member of a society in which all members receive publicly funded health care, as is the case in Canada, and in return contribute to improving the system. However, it remains fair to say that these are tentative efforts to move towards a new informed consent model and it falls to be seen whether and to what extent Canadian common law and research ethics guidelines will be flexible enough to adjust to biobank practices. Legislative initiatives may be required for such changes.

Focusing solely on informed consent seems, however, inappropriate for a variety of reasons. Biobank research raises complex legal, ethical, social and economic issues that cannot be adequately dealt with through informed consent procedures alone (see, for example, Laurie 2001).[4] One of the most fundamental issues is that much of the data contained in genetic samples and also some other health records are not unique to the person but are shared with and have implications for others. As the Icelandic Supreme Court recognized, people can have a privacy interest in the health information of a family member (*Gudmundsdóttir v. Iceland* [2003] No. 151). Many people will become incompetent or die during the time biobank research takes place. When donors of samples become incompetent or die, questions arise about the interpretation and even power of their initial consent or authorization for future use (Archibald and Lemmens 2008). It is difficult, in informed consent procedures, to account for all possible scenarios related to future use. Researchers will, for example, be interested in continued access to an incompetent person's clinical file, or may even want to conduct additional research related procedures involving the now incompetent person. Depending on the applicable legal rules, whether or not the original consent still pertains may not always be obvious, even when the person has signed an authorization for future use. Traditional informed consent procedures are also ill-equipped to deal with the interests of communities in some forms of biobank research. Controversies related to the use of samples of aboriginal communities and isolated communities, for example, highlight the need for other decision-making procedures and conflict resolution mechanisms, which do not rely solely on individual choice at the time of donation (see Wiwchar 2005,

4 Laurie points out that informed consent is often used to avoid key issues related to control, access, benefit sharing, and commercialization.

Austin, Harding and McElroy 2003, Mead 1996).[5] Issues of commercialization of biobank research findings and benefit sharing should also be mentioned at this stage. In this context, Gerald Porter refers to informed consent as a 'wolf in sheep's clothing', since informed consent disguises the commercial nature of many of the biobank transactions (Porter 2004: 85). The few American cases relating to ownership and control over biological samples confirm in our view the serious limits of relying on informed consent to deal with questions of benefit sharing, commercialization, and even directing the use of individual samples (*Moore v. Regents of the University of California* [1990] 51 Cal. 3d 120, *Greenberg* et al. *v. Miami Children's Hospital Research Institute* [2003] 264 2d 1064, *Washington University v. Catalona* [2007] 490 F.3d 667).

Practical Limits to Informed Consent: New Technological Developments

Informed consent as a protective mechanism is also under stress because of a series of technological developments, new research and data-sharing practices, and new research policies, which may even undermine the concept of anonymous data. Anonymization was until recently often seen as a necessary procedure to allow for more flexible consent procedures. As recently as 2003, a report by the World Health Organization suggested that blanket consent for future research was acceptable only 'in circumstances where anonymity of future data can be guaranteed'. '[I]f linking of data is necessary for the research', the report emphasizes, 'then specific consent to the research must be obtained' (Laurie 2004: 90). New developments in genetic technology combined with an ever-increasing availability of this information on various internet sites and publicly accessible databases seem to make this approach increasingly unworkable.

Personal genome scanning

The first technological development of note is the development of personal genome scanning. New genetic technology has significantly reduced the time and resources needed to sketch a map of an individual human genome. University and industry teams are currently competing for the $CAD10 million X-Prize Foundation's Genomic Prize, offered to the first team who decodes the DNA of 10 people in one week at a cost of less than $1,000 per genome. While it remains to be seen when this will become possible, the price of personal genome scanning

5 Wiwchar discusses the controversy surrounding the use of samples of the Nuu-chach-nulth, an aboriginal community of British Columbia, whose genetic samples were donated for use in arthritis research, but were later also used, without involvement of the community, in anthropological genetics. See also, Austin, Harding and McElroy's discussion of the controversy surrounding the development of a biobank in the Kingdom of Tonga and Te Pareake Mead, for a discussion of the way Maori people qualify genetic data.

has already been significantly reduced, opening the door to more widespread individual genome scanning.

Some of the promoters of individual genome scanning also embrace the idea that personal genomic information should be publicly shared. The Personal Genome Project of George Church, a Harvard biologist whose team is also participating in the X-Prize competition, is a case in point. This project involves the development of a database that will integrate detailed information on up to 100,000 people, including their individual genomes, medical records, imaging data (e.g. magnetic resonance imaging) and detailed personal information. The project claims to be 'proactive in dealing with one of the biggest problems with genomic medical research – the anonymity [of genomic data]' (Singer 2006). Its mission is to 'encourage the development of personal genomics technology and practices that … are broadly available for the good of the public' (http://personalgenomeproject.org: 22). Participants are asked to embrace a 'philosophy of absolute openness and sharing' (http://personalgenomeproject.org: 22).[6]

Not insignificantly, some of the ardent supporters and participants in this form of web-based genetic exhibitionism are also involved in other commercial data sharing ventures. Esther Dyson, for example, one of the participants, is an investor in *PatientsLikeMe.com*, a US based health care information company, as well as a board member of *23andMe*, a direct to consumer genetic testing company. Both companies frown upon traditional privacy concerns and want to radically depart from the idea that medical information should be kept confidential. In contrast to other health care sites that have a privacy policy, the website of the former states, 'at PatientsLikeMe, we're more excited about our Openness Philosophy'. This sharing of health information is given mythical power: it is associated with the company 'speeding up the pace of research and fixing a broken healthcare system'. Health care privacy regulations or 'proprietary tactics' are seen as a barrier to progress. People are encouraged to fully share information 'with other patients, caregivers, physicians, researchers, pharmaceutical and medical device companies, and anyone else that can help make patients' lives better'. *23andMe* also enthusiastically subscribes to the sharing philosophy that drives the projects described earlier. Although the company emphasizes that people should determine for themselves who has access to the information on their individual genome, it also embraces the value of sharing information with family members and interested members of the '23andMe' community and encourages people to participate in various specific disease or risk communities.

That this genetic exhibitionism creates informational risks for family members is hard to deny. Contrary to what Dyson confidently states in an op-ed article in which she explains her enthusiasm for putting her genome on the web – 'there are no deep secrets in my family' – we do not know what the sequencing of a person's

6 It emphasizes that it is looking for people 'interested in obtaining and openly sharing their genome sequences, related health and physical information, and reporting their experiences as a participant on the project'.

genome will reveal, either now, or in the future, when new risk-predictive tests become available.

Even when people are given an option to hide some details of their individual genome from public scrutiny prior to posting it on the internet, by removing certain genetic sequences or redacting the sequence (adding DNA letters to it), new genetic technology enables skilled geneticists to reconstruct the missing sequences. Indeed, Nyholt and his research team discuss step-by-step how a person's APOE genotype, associated *inter alia* with a risk for Alzheimer's disease, can be determined on the basis of an analysis of the surrounding gene sequences and information available in public databases (Nyholt, Yu and Visscher 2008, Callaway 2008). When Jim Watson, who had asked that genetic sequencing information related to potential Alzheimer's risk be removed from his personal genome prior to publication, was informed of the results of this study, he edited the area surrounding his APOE gene, by redacting 2 million DNA letters around it. Yet reconstruction remained possible, even if more cumbersome (Callaway 2008).

Data Sharing and Obligations to Share Research Results

Research related health information is also increasingly available as a result of the growing emphasis on the need for transparency and sharing of research data (Lemmens and Bouchard 2007). In the context of genetics, there is a particularly strong tradition of data sharing, going back to the efforts to map the human genome (see, for example, *The Human Genome Project Race* 2009). The 1996 Bermuda Rules state, among other things, that 'all human genomic sequence information, generated by centres funded for large-scale human sequencing, should be freely available and in the public domain in order to encourage research and development and to maximize its benefit to society' (*Summary of Principles Agreed at the First International Strategy Meeting on Human Genome Sequence* 1996, Marshall 2001, McGuire et al. 2008). And in 2003, the Wellcome Trust and the National Human Genome Research Institute, which are both funding major large-scale genomic sequencing, adopted a set of principles in Fort Lauderdale which confirm this commitment to fast and public release of sequencing data even prior to publication (*Sharing Data from Large Scale Biological Research Project* 2003). Other publicly funded initiatives, such as the Encyclopaedia Of DNA Elements Project (or ENCODE) of the National Human Genome Research Institute have also reaffirmed their commitment to quick data release and public accessibility of gene sequences.

This commitment to transparency and the sharing of data is a strategic component of public interest oriented science. Yet technological advances create new privacy concerns. In September 2008, the NIH announced it was withdrawing some sequencing data from publicly accessible databases. The NIH recognized that it has become possible to connect specific sequences in these databases to individual donors, if other identifiable genomic data is elsewhere available, thus undermining the anonymity of data (NIH 2008a, NIH 2008b).

This development shows the challenges we face in this new era, an era where improved gene sequencing technology and personalized genome testing meet individual genetic exhibitionism and public science oriented data sharing via publicly accessible websites. The protective gaze that quite solidly surrounded anonymized biological samples seems to have vanished, or at least been partially ruptured, undermining claims that informed consent procedures and individual control over our samples can offer us sufficient privacy protection in the context of genetics. The jury is still out on the extent to which biobank data could be connected somehow to anonymous sequencing data shared in research projects and to data shared by family members on health information sharing sites, and whether parties may be interested in using various data sources and new personal genome scanning technology to intrude upon the privacy of individuals. Yet the studies showing the possibility of connecting various data sets to de-anonymized data should constitute a warning sign. The anecdotal story of a 15-year-old boy who managed to find his biological sperm-donor father by linking very general information on the donor with his own genetic data and data provided by data-sharing websites, shows us how surprisingly easy those connections may have become (Motluk 2005).

Fair Information Practices and Governance of Biobanks

The discussion so far suggests that using informed consent as sole or primary regulatory mechanism for the use of data in biobanks has serious practical and conceptual limits. But in many countries, biobanks still operate without firm legislative frameworks, often for the practical reason that developing legislation takes time, and time is of the essence in the rush to develop biobanks. Timothy Caulfield has argued that biobank practice may in fact violate traditional legal consent requirements (2007) and recommends that a legislative framework be introduced for biobanks (2009). We agree with the argument that a coherent legislative basis for governing biobanks is desirable. In the remainder of the chapter, we want to point out, however, how several privacy statutes in Canada already contain the seeds of a different approach towards health information research, that is, one based on the concept of Fair Information Practices. Yet, we will also discuss how this approach requires a solid governance structure that respects basic requirements of good governance. Most of the existing privacy statutes in Canada rely in our view on governance bodies that are not capable of fulfilling these requirements. As we will point out at the end of this chapter, a new data governance structure introduced in British Columbia by legislation seems in this respect a commendable exception.

Fair information practices and Canadian privacy statutes

Fair Information Practices refers to a set of internationally accepted principles regarding the collection, use and disclosure of personal information. Their source is generally acknowledged as the *OECD Guidelines on the Protection of Privacy and Transborder Flows of Personal Data* (1980). While the protection of privacy is no doubt central to these Guidelines, the text refers more broadly to 'privacy and individual liberties' (*Guidelines on the Protection of Privacy* 1980: ss.2, 3(b) and 6). Eight major principles form the core of Fair Information Practices: Collection Limitation, Data Quality, Purpose Specification, Use Limitation, Security Safeguards, Openness, Individual Participation and Accountability (*Guidelines on the Protection of Privacy* 1980: Part Two). 'Fairness' in this context means accountability for the limited collection of relevant and accurate data that is used only for specified purposes and which is protected by adequate security. These practices should be transparent and individuals are entitled to know what information is being collected about them and challenge its accuracy.

There is no independent consent principle in the OECD Guidelines. Instead, consent plays a role in both the Collection Limitation Principle and the Use Limitation Principle. According to the Collection Limitation Principle, '[th]ere should be limits to the collection of personal data and any such data should be obtained by lawful and fair means and, where appropriate, with the knowledge or consent of the data subject' (*Guidelines on the Protection of Privacy* 1980: s.7). According to the Use Limitation Principle, personal data should not be used or disclosed except in accordance with the specified purposes unless (a) there is individual consent or (b) the use is authorized by law (*Guidelines on the Protection of Privacy* 1980: s.10). The OECD Guidelines therefore acknowledge that it may not always be appropriate to require consent for the collection of personal information and it may be possible to use this information for new purposes if such use is authorized by law.

In Canada many statutes are based upon Fair Information Practices and highlight both the important but also variable place of consent. For example, public sector data protection laws such as the federal *Privacy Act 1985* permit the government to collect personal information without consent, although the information must relate to a government program or activity, the government must generally collect this information directly from individuals, and it must inform them of the purpose of the collection (Privacy Act 1985: ss.4-5). In contrast, private sector data protection laws such as the federal *Personal Information Protection and Electronic Documents Act 2000* (PIPEDA), makes consent one of its central principles. However, despite this, PIPEDA also has many exceptions to the requirement of consent and permits a variety of means of obtaining consent, depending on circumstances (including implied, explicit, opt-in and opt-out) (Personal Information Protection and Electronic Documents Act 2000: s.7, Austin 2004). Therefore, despite the fact that informed consent has been called a 'fundamental pillar' of data protection law, this must be assessed with significant qualifications (Kosseim and Brady 2008).

First, consent is not always appropriate, as other values can take precedence over an individual's interests in their personal information. Second, even where consent is required the form of consent required varies. Third, the protection of privacy – and indeed other interests – requires much more than simply a focus on consent and includes a variety of security and accountability measures. The more stringent requirements for informed consent in relation to access to personal health information in the research context therefore come more from the research ethics paradigm than from the Fair Information Practices paradigm. As is widely known, the Nuremberg Code, the first truly internationally relevant promulgation of research ethics standards by the International Criminal Court in the Nuremberg Doctor Trials introduced the need for informed consent as an absolute requirement, in reaction to the horrific experiments on prisoners (*Trials of War Criminals before the Nuremberg Military Tribunals under Control Council Law* 1949: 181-182). Although subsequent guidelines and regulations, such as the Helsinki Declaration, introduced exceptions to this need for informed consent to allow research on children and incompetent adults (*World Medical Association Declaration of Helsinki* 1964, Lederer 2005), the norm in research ethics is clearly that informed consent is a core requirement for research participation. This difference in the emphasis on informed consent between privacy statutes and research ethics guidelines may create a particular problem when REB boards, used to operating within the context of research ethics guidelines, are asked to review research within the context of the Fair Information Practices paradigm. For the purpose of our discussion, it is interesting to look here at how two privacy statutes (Alberta and Ontario) contain provisions allowing research involving health data without informed consent if an REB has reviewed the research plan and has considered some other factors (see, for example, *Personal Health Information Protection Act 2004*: s.36 (1)(d), s.37 (3), s.44, *Health Information Act 2000*: s.48-56, s.72). Generally, according to both privacy statutes, REBs have to consider whether access to the data is essential to accomplish the research; that obtaining informed consent is impractical or unreasonable; and that appropriate safeguards are in place. In addition, both statutes also refer to the need for public interest considerations. Here, the language differs in a perhaps significant way. While the Ontario statute simply states that the REB has to consider the 'public interest in conducting the research and the public interest in protecting … privacy' (*Personal Health Information Protection Act 2004*: s.44 (3)(c)), Alberta REBs are asked to determine whether 'the proposed research is of *sufficient importance* that the public interest in the proposed research *outweighs to a substantial degree* the public interest in protecting the privacy of the individuals who are the subjects of the health information to be used in research' (emphasis added) (*Health Information Act 2000*: s.50 (1)(b)(i)). The Alberta act not only emphasizes that the public interest must be substantial, it also provides some guidance as to what these public interests are. It specifies that REBs should consider whether the research contributes to the following issues: the identification, prevention or treatment of illness or disease; a scientific understanding relating to health; the promotion and protection of the health of individuals and communities;

improved delivery of health services; and improvements in health system management (*Health Information Act 2000*: s.50 (2) (a-e)).

The provisions discussed here seem applicable in the context of biobank research. They do not reject the importance of some form of informed consent or initial authorization for storage of health information – informed consent is in fact still a key component of the regulatory regime in both statutes – but they reflect a realistic assessment that obtaining informed consent may, in the context of some important research endeavours, be impossible or very difficult.[7] Further, the Alberta Privacy Act gives us at least some idea of various types of research that can be considered of such public benefit that enabling the research can be more important than rigidly insisting on individual control over privacy. Finally, both the Ontario and Alberta statutes highlight the need for developing appropriate security measures in the context of research. We will say something more about that later.

REBs, Fair Information Privacy Practices and Good Governance Principles

At first sight, the reliance on REBs for governing Fair Information Practices seems reasonable. REBs have the mandate to protect the rights and wellbeing of subjects, including their privacy rights. Moreover, REBs are already accustomed – or so we should expect – to conducting a cautious evaluation of how research will affect various rights and what kind of risks and potential benefits are involved. REB review is characterized by a delicate exercise in interdisciplinary assessment of fair research practices. REBs must determine whether risks in research are 'reasonable', whether the risks are 'proportionate' to the expected benefits, whether there has been no 'undue' pressure to participate in research, whether there is a fair selection of research subjects, and so on.

Determining Fair Information Practices resembles the difficult balancing act REBs are supposed to engage in. The term itself indicates that we are talking here about the exercise of discretionary review. According to the privacy statutes, REBs have to determine whether research cannot 'reasonably' be accomplished without waiver of consent. They have to 'consider' the public interests in research and the public interest in protecting privacy. They have to determine whether obtaining consent is 'impractical' or 'unreasonable'. REBs are thus relied upon to exercise fairness and reflect a good balancing of the various interests in the use of health information, as they to some extent already do in the general context of research.

There are, however, some fundamental problems related to the reliance on this governance structure in the Canadian context. A discussion of some key principles of good governance helps to clarify why this is the case, and what we should expect of a governance structure in the context of health information related research, and

7 The term 'impractical' seems in our opinion too lenient. Obtaining informed consent is to some extent always impractical.

biobank research in particular. Governance is used here with the general meaning of the process of decision-making and the implementation and enforcement of these decisions. For the concept of 'good governance', it suffices here to refer to a couple of core, interrelated principles that can be found in many policy documents on good governance and to briefly elaborate why several of these criteria are not fulfilled in the context of the Canadian REB governance system. Good governance principles are: effectiveness and efficacy, accountability, respect for the rule of law, transparency, participation, and responsiveness.

The most significant problem is the lack of accountability. Accountability towards those affected by the decisions and in accordance with a stated governance objective or mandate is a core requirement of any governance system. Accountability also requires that governance systems are transparent and respect the principle of the rule of law, broadly conceived. At a minimum, decision-making has to be publicly verifiable as being in line with established rules and principles, and those who govern ought to be held responsible for their failure to adhere to the rules. In the context of research ethics governance, it is widely acknowledged that REBs have the mandate to ensure the protection of the rights and well being of research participants. REBs exist because of the recognized need for an external and independent evaluation of the possible impact of research on participants. Considering the focus of the REBs mandate, that is, the protection of rights and physical well-being, one would expect REBs to be accountable to the larger public and to be subject to governmental or quasi-public monitoring and enforcement of standards. Yet there is in most Canadian provinces no coherent regulatory framework surrounding REBs, no significant monitoring of their activities and no clear enforcement mechanism.

This is even more surprising since the public role of REB review is clearly recognized by federal funding agencies, by the federal drug regulatory agency, and by provincial governments. To begin with the latter: privacy statutes explicitly rely on REBs for governing Fair Information Practices in research, yet the statutes contain only limited provisions dealing with the organization, training, and public accountability of these REBs. The Ontario Privacy Act, for example, defines a Research Ethics Board simply as 'a board of persons that is established for the purpose of approving research plans under section 44 and that meets the prescribed requirements' (*Personal Health Information Protection Act 2004*: s.2 'research ethics board'). Except for a stipulation of core membership requirements, there are no further provisions related to the composition or functioning of these boards, to the need for terms of reference, quora, minute keeping, appeal process, and so on. There are no other provincial regulations dealing with REBs. The Alberta statute also does not specify any requirements in its *Health Information Act 2000*, but the Act provides the Minister of Health at least with the specific power to designate by regulations specific bodies as REBs for the purpose of the privacy statute (s.108 (2) (a)). A *Designation Regulation* (2001) has approved six existing REB with the important mandate under the *Health Information Act* to ensure Fair Information Practices in research.[8]

Few provinces have taken strong initiatives to structure REB governance (for example, *Plan d'action ministèriel en Éthique de la recherche et en intègritè scientifique* 1998).[9] In fact, the only province which recently provided a coherent legislative basis for REB review is the smaller province of Newfoundland (*Health Research Ethics Authority Act 2006*). Notwithstanding frequent calls for a coherent regulatory framework around REBs (Law Commission of Canada 2000, Lemmens 2005, Miller 2006, Downie 2006), neither the federal government nor most of the provincial governments have taken initiatives to comprehensively regulate REBs. Newfoundland is a noticeable exception, having introduced a statutory based review system.

As in the United States, there are two main sources of REB review in this country: the federal funding agencies and the regulations related to clinical research involving pharmaceuticals and medical devices (see, Lemmens 2005, Glass 2006, Hadskis 2007). Federal funding agencies require that all research that takes place in federally funded institutions has to abide by the Tri-Council Policy Statements, which contain detailed rules about REB review. Yet there is no official system of accreditation or certification of REBs and REB members, although there are proposals and plans to organize such an accreditation system through a not-for-profit organization (National Council on Ethics in Human Research 2006).

The federal *Food and Drugs Act 1985* and enabling Regulations (1985) also require REB review for all research that takes place in the context of the drug and medical device approval process. Sponsors of clinical trials for pharmaceutical products and medical devices must sign a form indicating that they have obtained the approval of a REB prior to obtaining regulatory approval to use a non-approved product in a clinical trial. But there is no official approval system and no detailed oversight system for REBs. The regulatory authorities have, for example, no official count of the REBs operating in the country. A REB can in principle be set up overnight and still satisfy the criteria for drug approval. Health Canada, the agency responsible for implementing the *Food and Drugs Act* and Regulations, has implemented the International Conference on Harmonization of Good Clinical Practice Guidelines (ICH-GCP) as a guidance document in its regulatory system (*Guidance for Industry*). This document does not have the status

8 Unlike Ontario, Canada's most populated province with most of the country's leading genomics and other research institutes, Alberta has thus a clearly established list of REBs that can review research for the purposes of privacy legislation. But both provinces rely on the existence of other norms and other regulatory control on REB governance.

9 Newfoundland, Alberta, and Quebec can probably be cited as the provinces with at least more detailed regulatory initiatives related to research ethics. The province of Quebec has developed a detailed ministerial plan for research ethics, which includes oversight of provincially funded hospital and institutional REBs and a comprehensive educational plan.

of strict regulations and is thus harder to enforce (Lemmens 2005). Moreover, the ICH-GCP relies on national regulations for more detailed provisions with, for example, the functioning and authority of REBs (Hirtle, Lemmens and Sprumont 2000). The monitoring process for REBs is also very limited. Health Canada has started to conduct a limited number of site visits of Canadian REBs as part of its authority over the control of clinical trials, but it is stretched in its resources for this monitoring (Lemmens 2005).

Finally, it is worth pointing out that research that takes place outside federally funded research institutes, and that does not involve the testing of drugs and medical devices for regulatory approval, seems to escape any requirement related to research ethics review. While most large-scale biobank research takes place with federal funding and in federally funded institutions, and while provincial statutes put restrictions on the storage of human tissue (see, for example, *Trillium Gift of Life Network Act 1990*), commercial research related to genetic samples outside of funded institutions and outside the drug evaluation context is not subject to any of the regulatory regimes with research ethics requirements. If this research involves activities that fall under health privacy legislation, privacy provisions related to research, including REB review requirements, could still apply. Yet these REBs would have no regulatory or even soft-law or contractual basis (that is, they would in theory neither be bound by the ICH-GCP guidance document of Health Canada, nor by the Tri-Council Policy Statement of the Funding Agencies). This clearly does not satisfy even the weakest standards of accountability.

In most provinces, including Ontario, REBs do not have a clearly established territorial jurisdiction, which creates opportunities for forum shopping. Many private REBs operate in the country and compete for research ethics review business, particularly in the context of clinical drug trials. Some of these REBs are cloaked in secrecy, refusing for example to reveal the identity of their members, a remarkable phenomenon for an administrative body with a public mandate. Private commercial REBs may be the REBs of choice for commercially sponsored biobanks set up by biotechnology or pharmaceutical companies. Nothing prevents sponsors of research taking place outside of academic institutions to shop for a REB that is more lenient, or more flexible with respect to some requirements. Biobank research will generally be connected to major research institutes with designated REBs, but forum shopping between institutions is also possible. When institutional REBs create difficulties for certain forms of research, the institution may lose potentially lucrative research contracts or funding. Members of institutional REBs may feel under pressure when influential colleagues submit a research project that will bring in significant prestige and research funding to the institution. Conflicts of interest exist thus for both commercial REBs (who compete for business), and institutional REBs (who may fear that their actions can impact on the research resources within the institution).

The structural conflict of interest that surround Canadian REBs also affects their ability to be truly responsive to one of the core issues in biobank governance, that is, the potential tension between commercial interests of research and the public interest or the interest of individual research subjects. Indeed, many of the admittedly few legal challenges involving human tissue research are related to the commercial exploitation of research results (see, for example, *Washington University v. Catalona* (2007) 490 F.3d 667, *Greenberg et al. v. Miami Children's Hospital Research Institute* (2003) 264 2d 1064, *Moore v. Regents of the University of California* (1990) 51 Cal. 3d 120). Many academic commentators emphasize the need to develop a framework for the governance of benefit sharing in the context of biobank research (see Winickoff 2008, Bovenberg 2005, Bouchard and Lemmens 2008). If, as the concept of Fair Information Practices suggest, REBs are to weigh *inter alia* the interests of individuals and the public interest in research, as they are asked to do under the privacy statutes mentioned earlier, they must be sufficiently independent from the commercial sources that may affect both interests.

Canadian REBs sometimes explicitly conflate their role in protecting human research subjects with a role in promoting research and in the commercialization of medical research.[10] Two recent publications emanating from the Ontario Cancer Research Ethics Board (OCREB) provide evidence of this conflation of roles within the academic context. OCREB is an influential Ontario multicentre REB, which also has a Tissue Ethics and Scientific Committee for evaluation of access to samples and biobank related data. This REB operates within the Ontario Institute of Cancer Research, a well-regarded and highly successful not-for-profit corporation funded by the Ontario Government. In one recent publication, the REB is praised for its 'ability ... to shorten and coordinate study start-up times [which] imparts a competitive edge to Ontario, not only making the province an attractive location in which to conduct cancer clinical trials, but also improving access to novel therapies' (Chaddah 2008). Another article emphasizes how the quality, efficiency and timeliness of review help 'maintain the trust of sponsors, investigators, and institutions'. Explicit reference is also made to the importance of 'enhancing human participant protection' but in combination with the need to make 'Ontario a more attractive region to conduct cancer clinical trials' (Saginur et al. 2008). There is obviously nothing wrong with promoting clinical research activities and promoting efficient research ethics review. This seems also to be in line with the governmental mandate of the umbrella organization.[11] But it is

10 While the interests of research subjects and research itself can coincide, it has to be kept in mind that REB review has been developed precisely because many historical controversies provide evidence of how research interests and commercial interests can clash with the public interest and the interests of research subjects.

11 The Ontario Institute of Cancer Research, under which the REB operates, clearly has a commercialization mandate. Reference is made to the need to create 'a culture that recognizes the value of translation and commercializing cancer-related innovations in bringing the benefits of discovery to Ontario residents'. Much emphasis is placed on the importance of creating intellectual property rights, and to the need to 'promote the downstream development of inventions and ultimately the manufacturing and production of devices derived from these inventions in Ontario'.

perhaps paradigmatic of the Canadian REB scene that an administrative body that ought to have as its core mandate the protection of research subjects and promotion of ethics standards, feels the need to portray its services as part of a mission to stimulate research investments and innovation.[12] And it may also reflect how REBs in the country compete for business and try to convince researchers, institutions, and sponsors, to work with them rather than with another REB.

The lack of clear regulatory requirements and the presence of potential conflicts of interest are particularly problematic because of the nature of research ethics review in general, and the nature of establishing fair research information practices in particular. First, as mentioned earlier, research ethics review is clearly an exercise of a public mandate. It aims to protect the well-being of participants in research and to ensure their rights are respected. It also aims at preventing conflicts of interest affecting the conduct of research. This type of protective mandate is usually exercised by administrative or judicial bodies that are fully independent from those with interests that may clash with those who are to be protected. Second, as pointed out, both REB review and implementation of Fair Information Practices require a cautious assessment of the various interests at stake. We necessarily rely on the good judgment and fairness of the decision makers in this process. While private entities can be empowered with a public mandate, rules about independence, conflict of interest, and public accountability become more important when the nature of the administrative activity is such that the content of substantive rules is vague, and the administrative body has to exercise good judgment and reflect fair decision making (Lemmens and Freedman 2000).

The effectiveness and efficiency requirements of good governance also pose problems for the Canadian REB system. Many REBs have been criticized by the research community for their lack of efficiency. They are generally staffed by volunteer members. Many are overwhelmed by the number of applications and lack sufficient administrative support. REBs may also face particular problems in implementing effective privacy protection. The concept of Fair Information Practices is relatively new in the Canadian research ethics scene and there is a lack of privacy expertise and training of REB members in this area. REBs are generally accustomed to reviewing informed consent procedures, looking at payment structures, stipulating required disclosures of conflicts of interest, requesting data monitoring, verifying administrative requirements, and some may also be doing a reasonable job in weighing risks and potential benefits in more traditional medical research, such as clinical drug trials. Until recently, however, they were far less

12 It is appropriate to disclose here that one of us was consulted at the time of the establishment of the REB about the administrative structure and organization of the REB and explicitly recommended (submission prepared with Duff R. Waring) on the basis of conflict of interest considerations that the REB report to the administration of Cancer Care Ontario, a clinical care branch of this Ontario cancer care network, and not to the administrators responsible for attracting and promoting research.

frequently confronted with issues related to informational risks, such as those raised in the context of biobank research.

To take one example, it is clear from our previous discussion of Fair Information Practices that the protection of privacy contemplates much more than simply obtaining individual consent. One extremely important principle is security of the information: parties who collect, use or disclose personal information should be required to maintain adequate security in order to prevent unauthorized disclosures. Indeed, in both the Ontario and Alberta statutes REBs are to consider whether adequate safeguards are in place to protect the privacy and confidentiality of the health information in question (see, *Health Information Act 2000*: s.50 (1)(b)(iii), *Personal Health Information Protection Act 2004*: s.44 (3)). However, to do so requires specialized knowledge regarding information systems and anonymization protocols and it is unclear whether REB members possess that expertise. It is interesting to refer here again to the *Gudmundsdóttir* (2003) case, in which the Icelandic Supreme Court emphasized the need for a solid regulatory framework surrounding biobanks. 'Owing to the [constitutional] obligations [to ensure privacy protection]', the Supreme Court stated, 'this assurance cannot be replaced by various forms of monitoring of the creation and operation of the Health Sector Database, monitoring which is entrusted to public agencies and committees without definite statutory norms on which to base their work' (*Gudmundsdóttir v. Iceland* [2003] No. 151: 9). In Iceland, a legislative framework for the public agencies and committees involved in the monitoring process was in place at the time of this decision, but was thought to be too vague. What is there to say then about the Canadian reliance on administrative bodies that lack a clear regulatory structure, are not subject to rigorous oversight, are not sufficiently accountable to public authorities, have no clear educational and training requirements, have conflicting interests, and compete for business? Clearly, Canadian research ethics governance seems to fail the most basic test of public accountability.

Models of Good Governance of Biobanks

Let us conclude, however, with some good news from Canada. In addition to the laudable initiatives in some provinces to improve REB governance discussed earlier, the province of British Columbia has a new *Personal Health Information Access and Protection of Privacy Act 2008*, which introduces a governance model that seems to afford much greater respect to the good governance requirements discussed here. The Act empowers the Minister of Health to set up health information banks for the purpose of conducting or facilitating research. Data Stewardship Committees will be set up in connection with these information banks, with the responsibility to manage the disclosure, research planning, and sharing of information in these banks. The Committee will consist of members with an interested and very relevant range of expertise or representative status. The statutes require representation from the Ministry of Health, the Regional

Health Boards or Health Authorities, the research community, and the Doctors, Pharmacists, and Nurses organizations. In addition, the Act requires that there be 3 representatives from the general public.

This seems to be a very good model for the implementation of Fair Information Practices in the context of biobank research. First, a clear legislative authority, with a clear public accountability mechanism is built into the statute. Second, the membership of this committee is more balanced and in line with the mandate of a governance structure that has to focus on the protection of the interests of those whose information will be used, the public interest in the research, and the public interest in privacy. Governmental agencies that have as their official mandate the promotion of the health of the citizens are given an appropriate voice here. In addition, the inclusion of at least three members of the public provides a more solid basis for real and meaningful community input. It will help empower those who, in the current REB context, where only one community member is required, are often voiceless and isolated. Stronger community involvement is sometimes resisted by those who are worried that it may inappropriately slow down research, because of the lack expertise or understanding of often highly technical and complex research procedures. This is not a valid reason to resist better community involvement. First of all, if community involvement is important, we have to accept that it may come at a price. And importantly, there are sufficient models out there of research ethics review or research governance structures with very substantial public involvement, which does not seem to compromise the effectiveness and efficiency of the system. The Ontario HIV Treatment Network Cohort Study is a case in point (Ontario HIV Treatment Network 2008). This study involves a very extensive database of all clinical data of people suffering from HIV/AIDS in the province. Access to the database for research purposes is determined by Governance and Scientific Steering Committees, half of the members of which are recruited from the affected community. Researchers involved in the project are enthusiastic about the model, which seems to improve cooperation and active involvement of the community in the research endeavours.

This system needs further development and implementation. Emphasis has to be placed on the need for appropriate training of data committee members in privacy measures. Privacy specialists ought to be included in the governance committees associated with biobank research. The data stewardship committees must receive the necessary regulatory and budgetary means to implement a coherent monitoring and enforcement system in order to ensure compliance. Attention will also have to be paid to the development of clearer guidelines to assist in the determination of what constitutes the public interest in research and in the weighing of this interest against the interests of individuals and the public interest in the protection of privacy.

Governments must invest in the development of good governance structures for biobanks, not out of mistrust for the research that will take place, but out of respect for the public health importance that these research infrastructures represent, and out of respect for the committed individuals who are inevitably giving up some of their privacy to promote a public good.

Bibliography

23andMe. [Online] Available at: https://www.23andme.com/health/ [accessed 16 December 2008].

Archibald, T. and Lemmens, T. 2008. Data Collection from Legally Incompetent Subjects: A Paradigm Legal and Ethical Challenge for Population Databanks. *Health Law Journal* (Special Edition). [Online] Available at: http://ssrn.com/abstract=1267204.

Austin, L.M. 2004. Is Consent the Foundation of Fair Information Practices? Canada's Experience Under PIPEDA. *University of Toronto Law Journal*, 54, 183.

Austin, M.A., Harding, S. and McElroy, C. 2003. Genebanks: A Comparison of Eight Proposed International Genetic Databases. *Community Genetics*, 6, 37.

Bouchard, R. and Lemmens, T. 2008. Privatizing Biomedical Research – A 'Third Way'. *Nature Biotechnology*, 1, 31.

Bovenberg, J. 2005. Whose Tissue is it Anyway? *Nature Biotechnology*, 23, 929.

Callaway, E. 2008. DNA Detectives Can DeCode 'Censored' Genomes. *Short Sharp Science New Scientist News Blog*. [Online] Available at: http://www.newscientist.com/blogs/shortsharpscience/2008/10/dna-detectives-can-decode-cens.html.

Caulfield, T. 2002. Gene Banks and Blanket Consent. *Nature Reviews Genetics*, 3, 2.

Caulfield, T. 2007. Biobanks and Blanket Consent: The Proper Place of the Public Good and Public Perception Rationales. *Kings Law Journal*, 18, 209.

Caulfield, T.R., Upshur, R. and Daar, A. 2003. DNA Databanks and Consent: A Suggested Policy Option Involving an Authorization Model. *BMC Medical Ethics*, 4(1).

Chaddah, M.R. 2008. The Ontario Cancer Research Ethics Board: A Central REB that Works. *Current Oncology*, 15, 49.

Deschênes, M. et al. 2001. Human Genetics Research, DNA Banking and Consent: A Question of Form. *Clinical Genetics*, 59, 221.

DNA Detectives Can DeCode 'Censured' Genomes. 2008. [Online: New Scientist Science Blog] Available at: http://www.newscientist.com/blogs/shortsharpscience/2008/10/dna-detectives-can-decode-cens.htm [accessed 1 January 2009].

Downie J. 2006. The Canadian Agency for the Oversight of Research Involving Humans: A Reform Proposal. *Accountability in Research*, 13, 75.

ENCODE Consortia Data Release, Data Use, and Publication. 2008. [Online: FlyBase] Available at: http://flybase.org/static_pages/news/articles/2008_09/encode_data_release _policy.pdf.

Glass K.C. 2006. Questions and Challenges in the Governance of Research Involving Humans: A Canadian Perspective, in *Law and Ethics in Biomedical Research: Regulation, Conflict of Interest, and Liability*, edited by Trudo Lemmens and Duff R. Waring. Toronto: University of Toronto Press, 35.

Godard, B. et al. 2003. Data Storage and DNA Banking for Biomedical Research: Informed Consent, Confidentiality, Quality Issues, Ownership, Return of Benefits. A Professional Perspective. *European Journal Human Genetics*, 11, 88.

Greely, H.T. 2001. Informed Consent and Other Ethical Issues in Human Population Genetics. *Annual Review of Genetics*, 35, 785.

Guidelines on the Protection of Privacy and Transborder Flows of Personal Data. 1980. [Online: Organization for Economic Co-operation and Development] Available at: http://www.oecd.org/document/18/0,3343,en_2649_34255_1815186_1_1_1_1, 00.html [accessed 29 December 2008].

Hadskis, M. 2007. The Regulation of Human Biomedical Research, in *Canada*, in *Canadian Health Law and Policy.* 3rd Edition, edited by J. Downie, T. Caulfield and C. Flood. Canada: LexisNexis.

Health Canada. 1996. *Guidance for Industry – Good Clinical Practice: Consolidated Guideline*. [Online] Available at: http://www.hc-sc.gc.ca/dhp-mps/alt_formats/hpfb-dgpsa/pdf/prodpharma/e6-eng.pdf [accessed 29 December 2008].

Hirtle, M., Lemmens, T. and Sprumont, D. 2000. A Comparative Analysis of Research Ethics Review Mechanisms and the ICH Good Clinical Practice Guideline. *European Journal of Health Law*, 7, 265.

Kaye, J. 2004. Broad Consent – the Only Option for Population Genetic Databases? in *Blood and Data, Ethical, Legal and Social Aspects of Human Genetic Databases*, edited by G. Árnason, S. Nordal and V. Árnason. Reykjavík: University of Iceland Press, 103.

Kosseim, P. and Brady, M. 2008. Policy by Procrastination: Secondary Use of Electronic Health Records for Health Research Purposes. *McGill Journal of Law and Health*, 2, 5.

Laurie, G. 2001. *Genetics Privacy: A Challenge to Medico-Legal Norms*. Cambridge: Cambridge University Press.

Laurie, G. 2004. Genetic Databases: Assessing the Benefits and Impact on Human and Patient Rights – A WHO Report. *European Journal of Health Law*, 11, 87.

Law Commission of Canada, *The Governance of Health Research Involving Human Subjects*. Ottawa: October 2000.

Lederer, S.E. 2005. Research without Borders: The Origins of the Declaration of Helsinki, in *History and Theory of Human Experimentation*, edited by U. Schmidt and A. Frewer. Stuttgart: Franz Steiner Verlag, 145.

Lemmens, T. 2005. Federal Regulation of REB Review of Clinical Trials: A Modest but Easy Step Towards an Accountable REB Structure in Canada. *Health Law Review*, 13.

Lemmens, T. and Bouchard, R. 2007. Mandatory Clinical Trial Registration: Rebuilding Trust in Medical Research, in Global Forum for Health Research,

Global Forum Update on Research for Health, Volume 4: Equitable Access: Research Challenges for Health In Developing Countries. London: Pro-Book Publishing, 40-46.

Lemmens, T. and Freedman, B. 2000. Ethics Review for Sale? Conflict of Interest and Commercial Research Review Boards. *The Milbank Quarterly*, 78, 547.

Levy, S. et al. 2007. A New Human Genome Sequence Paves the Way for Individualized Genomics. *PLoS Biology*, 5, e254.

Luther, L. and Lemmens, T. 2007. Human Genetic Databanks: From Consent to Commercialization – An Overview of Current Concerns and Conundrums, in *Biotechnology, Encyclopedia of Life Support Systems*, edited by W. Horst Doelle and E.J. DaSilva.

Marshall, E. 2001. Bermuda Rules: Community Spirit, With Teeth. *Science*, 291, 1192.

McGuire, AL. and Gibbs, RA. 2006. No Longer De-Identified. *Science*, 312, 370-371.

McGuire, A.L. et al. 2008. Ethical, Legal, and Social Considerations in Conducting the Human Microbiome Project. *Genome Research*, 18, 1861.

Miller, P.B. 2006. Institutional Oversight of Clinical Trials and the Drug Approval Process. *Osgoode Hall Law Journal*, 44(4), 679-726.

Motluk, A. 2005. Anonymous Sperm Donor Traced on Internet. *New Scientist*, 2524, 6.

National Council on Ethics in Human Research, Task Force for the Development of an Accreditation System for Human Research Protection Programs. 2006. *Promoting Ethical Research with Humans*. [Online: National Council on Ethics in Human Research] Available at: http://ncehr-cnerh.org/english/task_force [accessed 1 January 2009].

National Bioethics Advisory Commission. 1999. *Research Involving Human Biological Materials*. [Online: National Bioethics Advisory Commission] Available at: http:// bioethics.georgetown.edu/nbac/hbm.pdf [accessed 29 December 2008].

NIH. 2008a. *Background Fact Sheet on GWAS Policy Update*. [Online: National Institutes of Health] Available at: http://grants.nih.gov/grants/gwas/background_fact_sheet_ 20080828.pdf [accessed 18 September 2008].

NIH. 2008b. *Modifications to Genome-Wide Association Studies (GWAS) Data Access*. [Online: National Institutes of Health] Available at: http://grants.nih.gov/grants/gwas/ data_sharing_policy_modifications_20080828.pdf [accessed 18 September 2008].

Nyholt, D.R., Yu, C.E. and Visscher, P.M. 2008. On Jim Watson's APOE Status: Genetic Information is Hard to Hide. *European Journal of Human Genetics*. [Online] Available at: http://www.nature.com/ejhg/journal/vaop/ncurrent/abs/ejhg2008198a.html [accessed 1 January 2009].

Ontario HIV Treatment Network. 2008. *What is the OHTN Cohort Study?* [Online] Available at: http://www.ohtn.on.ca/OHTNCohort.htm [accessed 24 March 2009].

Patientslikeme. [Online] Available at: www.patientslikeme.com [accessed 26 December 2008].

Personal Genomics Project. [Online] Available at: http://www.personalgenomes.org/mission.html [accessed 22 December 2008].

Plan d'action ministériel en Éthique de la recherche et en intègrité scientifique. 1998. [Online: Sante et Services sociaux] Available at: http://publications.msss.gouvqc.ca/acrobat/f/documentation/1998/98_759.pdf [accessed 30 December 2008].

Porter, G. 2004. The Wolf in Sheep's Clothing: Informed Consent Forms as Commercial Contracts, in *Blood and Data, Ethical, Legal and Social Aspects of Human Genetic Databases*, edited by G. Árnason., S. Nordal and V. Árnason. Reykjavík: University of Iceland Press, 85.

PRE. 2003. *Tri-Council Policy Statement: Ethical Conduct for Research Involving Humans*. [Online] Available at: http://www.pre.ethics.gc.ca/english/pdf/TCPS %20June 2003_E.pdf [accessed 29 December 2008].

Saginur, R. et al. 2008. Ontario Cancer Research Ethics Board: Lessons Learned From Developing a Multicenter Regional Institutional Review Board. *Journal of Clinical Oncology*, 26, 1479.

Sharing Data from Large-scale Biological Research Project: A System of Tripartite Responsibility: Report of a meeting organized by the Wellcome Trust and held on 14-15 January 2003 at Fort Lauderdale, USA. 2003. [Online: National Human Genome Research Institute] Available at: www.genome.gov/Pages/Research/Wellcome Report0303.pdf.

Singer, E. 2006. *The Personal Genome Project*. [Online: Technology Review] Available at: http://www.technologyreview.com/BioTech/wtr_16169,259,p1.html.

Summary of Principles Agreed at the First International Strategy Meeting on Human Genome Sequencing. 1996. [Online: Human Genome Project Information] Available at: http://www.ornl.gov/sci/techresources/Human_Genome/research/bermuda.shtml [accessed 16 December 2008].

Te Pareake Mead, A. 1996. Genealogy, Sacredness, and the Commodities Market. *Cultural Survival Quarterly*, 46.

The ENCODE Project: ENCylopedia of DNA Elements. [Online: National Human Genome Research Institute] Available at: http://128.231.210.112/10005107#1.

The Human Genome Project Race. [Online: Center for Biomolecular Science and Engineering] Available at: http://www.cbse.ucsc.edu/research/research_hgprace.shtml [accessed 29 December 2008].

Trials of War Criminals before the Nuremberg Military Tribunals under Control Council Law. 1949. Washington, DC: US Government Printing Office, 181-182.

Wadman, M. 2008. James Watson's Genome Sequenced at High Speed. *Nature*, 452 (7189), 788.

Wheeler, D.A. et al. 2008. The Complete Genome of an Individual by Massively Parallel DNA Sequencing. *Nature*, 452, 872.

Winickoff, D. 2008. *From Benefit Sharing to Power Sharing: Partnership Governance in Population Genomics Research*, Working Paper No. 68, Berkeley: University of California.

Wiwchar, D. 2005. Long Journey Home Just the Beginning. *Windspeaker*, 22, 13.

WMA. 1964. *World Medical Association Declaration of Helsinki: Ethical Principles for Medical Research Involving Human Subjects.* [Online: World Medical Association] Available at: http://www.wma.net/e/policy/b3.htm [accessed 29 December 2008].

Wright Clayton, E. et al. 1995. Informed Consent for Genetic Research on Stored Tissue Sample. *Journal of American Medical Association*, 274 (22), 1786.

Legislation

Designation Regulation (Health Information Act) 2001. SI 2001/69, Alberta: AG.

E-Health (Personal Health Information Access and Protection of Privacy) Act 2008. (c.38), British Columbia: BCG.

Food and Drugs Act 1985. (c. F-27), Ottawa, Ontario: CG.

Food and Drug Regulations 1985. SI 1985/c. 870, Ottawa, Ontario: CG.

Health Information Act 2000. (c. H-5), Alberta: AG.

Health Research Ethics Authority Act 2006. (c. H-1.2.), St. John, Newfoundland and Labrador: LG.

Personal Health Information Protection Act 2004. (c.3), Ottawa, Ontario: CG.

Personal Information Protection and Electronic Documents Act 2000. (c.5), Ottawa, Ontario: CG.

Privacy Act 1985. (c.P-21), Ottawa, Ontario: CG.

Trillium Gift of Life Network Act 1990. (c. H. 20), Ontario: OG.

Legal authorities

Greenberg et al. *v. Miami Children's Hospital Research Institute* [2003] 264 2d 1064.

Gudmundsdóttir v. Iceland (2003) No. 151/2003.

Halushka v. University of Saskatchewan [1965] 52 WWR 608.

Moore v. Regents of the University of California (1990) 51 Cal. 3d 120.

Reibl v. Hughes (1980) 2 S.C.R. 880.

Washington University v. Catalona (2007) 490 F.3d 667 (8th Cir).

Weiss v. Solomon [1989] A.Q. No. 312.

Chapter 17

From Public Inquiry to Policy: Biobanks, Population Genetics and the Public Interest

David Weisbrot

In response to growing concern within Australian society, the Australian Law Reform Commission (ALRC) has completed two major inquiries dealing directly with the ethical, legal and social implications of human genetic research and the governance of biobanks. The first, in 2003, was concerned primarily with the use of genetic information (Genetic Information Enquiry) and resulted in a report called *Essentially Yours: The Protection of Human Genetic Information in Australia* (*Essentially Yours*).[1] The second inquiry in 2008, concerned the adequacy of Australian privacy laws and practices[2] (Privacy Inquiry) and resulted in a report called *For Your Information: Australian Privacy Law and Practice* (*For Your Information*). From the extensive community consultation process that was involved in these inquiries, the ALRC has formulated a number of policy recommendations. This process provides an insight into the relationship between public engagement and policy, and how the findings from community consultation exercises can provide a platform for the development of recommendations. This chapter will discuss the current situation in regard to biobanking in Australia; the nature of the community consultation process carried out by the ALRC; the findings of the ALRC based on the extensive public consultation processes undertaken for both these inquiries; and the ALRC policy recommendations that flowed from these findings.

---

1 The 2003 ALRC report, *Essentially Yours: The Protection of Human Genetic Information in Australia* (hereafter, Essentially Yours) made 144 recommendations for reform; available at: http://www.austlii.edu.au/au/other/alrc/publications/reports/96/.

2 The 2008 ALRC report, *For Your Information: Australian Privacy Law and Practice* (hereafter, For Your Information) contained 295 recommendations for reform. Part H of the report deals with privacy regulation in the specific contexts of health and medical information (Chapters 60-63) and research (Chapters 64-66), including databases and data linkages (Chapter 66); available at: http://www.austlii.edu.au/au/other/alrc/publications/reports/108/.

Australian Biobanking

Australia does not presently have – nor are there imminent plans to construct – a comprehensive or national biobank, such as has been established in Iceland (deCODE), Estonia (the Estonian Genome Project), the United Kingdom (UK Biobank) and Japan, and foreshadowed in Taiwan. However, there are state-wide projects in early development in Western Australia and in Tasmania, which involve collection of samples from volunteers on a large-scale basis, as well as extensive data linkage wherever possible. In addition, there numerous collections through the health service that are not co-ordinated or linked up in any systematic way.

In Tasmania, the Menzies Research Institute operates a research database comprising extensive genealogical data, and genetic samples and health information donated by volunteers specifically for the Institute's research projects.[3] In August 2006, the University of Western Australia Institute for Medical Research (WAIMR) announced that it was leading a team proposing to create a state-wide biobank (Skatssoon 2006. See also the WAIMR website at http://www.genepi.org.au) with links to the Public Population Project in Genomics (P^3G), an 'international consortium for the development and management of a multidisciplinary infrastructure for comparing and merging results from population genomic studies'.[4] The WAIMR initiative expressly recognized the critical importance of safeguarding privacy, calling for 'national privacy legislation and beefed up genetic data protection laws, in line with the recommendations of the Australian Law Reform Commission's *Essentially Yours* report' (Skatssoon 2006).

There was also a highly publicized, but ultimately unsuccessful, attempt to establish an Australian-controlled regional biobank. In November 2000, the biotechnology company Autogen informed the Australian Stock Exchange that it had signed an agreement with the Pacific Island nation of Tonga, to establish a major health database to identify genes that cause common diseases. It was noted that the 'unique family structure and isolation of this population together with the high prevalence of a variety of diseases represents a major resource for geneticists' (ABC News 2000, Howe 2001). Media reports at the time also talked of Autogen looking to expand this initiative to include one or more other countries overseas, as well as to Tasmania (ABC News 2000, Howe 2001). Although Autogen's statement on ethics emphasized the prior informed consent of individual Tongan volunteers, and the company promised to provide much-needed healthcare facilities, the project soon ran into an international storm of protest and did not proceed. Among

---

3 The Centre is affiliated with the University of Tasmania; see its homepage available at: www.menzies.utas.edu.au/.

4 See the P3G website: http://www.p3gconsortium.org/about.cfm; the founding members of the consortium include CartaGene, the Estonian Human Genome Project and an eight country EU-based biobank – but there now also other members from the US (the National Institutes of Health and the Center for Disease Control), Canada, Spain, the UK, Germany, Sweden, and a Central European group.

other things, there were serious concerns raised within Tonga and outside about: the secrecy of the negotiations, with a poor and vulnerable island state that is the region's only remaining full monarchy; the potential for 'bio-piracy', with the islanders' genetic resources exploited without fair value; and the possibility of obtaining truly informed consent from individual Tongans, given the powerful and hierarchical communal social structure (Burton 2002).

Apart from the budding biobanks in Tasmania and WA, there are a large number of smaller or unsystematized collections in Australia, maintained by universities, research centres and biotechnology companies; public and private hospitals (for example, tissue banks, blood banks, pathology samples, paraffin blocks); pathology labs; and familial cancer registers (ALRC 2003: Chapters 18-20 and 24). Perhaps of most potential importance, there exists a major *inchoate* national biobank in Australia, in the form of vast numbers of newborn screening blood spot cards (aka 'Guthrie cards'), stored in children's hospitals around the country (with varying degrees of care) – representing a complete, if unorganized, DNA collection of almost everyone born in Australia since about 1960. Guthrie cards are regarded as 'health records', and thus issues relating to storage, access and disclosure are generally governed by privacy legislation.[5] The Australian National Pathology Advisory Committee Guidelines recommend retention of the cards for 25 years, but there is no active program of removing or destroying older cards, while the policy in one state (Western Australia) is to destroy the cards after two years.

Public Engagement by the ALRC

It is normal practice for the ALRC to endeavour to engage the general public, as well as experts and interest groups, in its work – although the subject matter of some inquiries lends itself better to widespread community consultation than others. In both inquiries, the ALRC went to special lengths to ensure broad community participation. Apart from the production of the standard written public consultation documents, the ALRC also produced shorter, more accessible summaries and pamphlets, and promoted a great deal of coverage in the mass media. Across the two inquiries, the ALRC conducted nearly 500 face-to-face public forums, roundtables, workshops and 'targeted' meetings with key stakeholders and community organizations in Australia and overseas (about 225 for Genetic Information, and 250 for Privacy), and received nearly 1,000 formal written submissions (about 350 for Genetic Information, and 600 for Privacy). The ALRC ran a well-publicised National Privacy Hotline in June 2006, and developed a dedicated 'Talking Privacy' portal on its website aimed at encouraging interest

5 However, it is less clear what the status is of the blood sample once it is removed from the card – the blood/DNA no longer constitutes a 'health record', and would no longer be governed by the *Human Tissue Acts* (or, apparently, any other current law).

and participation by children and young people. In all, many tens of thousands of Australians engaged directly with either or both of these inquiries.

The ALRC's community consultation exercise confirmed the findings from other local and overseas surveys (Alboim 2005: 23-27, Eurobarometer 2003) and literature in respect of social attitudes to human genetic research and medicine; that is, there are strong, but conflicting, feelings in the community about biotechnology. There is considerable optimism about the potential for genetic research to produce important medical breakthroughs in the diagnosis, treatment and prevention of debilitating diseases, such as diabetes, Alzheimer's and Parkinson's, as well as leading to the development of whole new fields of medicine, such as gene therapy, regenerative medicine, and pharmacogenomics. There is also strong support in the general community for the use (with attendant safeguards) of DNA technology, analysis and databases by law enforcement and national security authorities.

At the same time, there is a palpable, underlying anxiety in the community about the pace of change – concerns about the loss of control; fears about the beginnings of 'genetic determinism' or perhaps even more radical eugenics; and some doubts about the capacity of public authorities to regulate this area effectively in the public interest, with regard to privacy protection, non-discrimination and stigmatization, and ethical oversight of human genetic research. Put another way, Australians are attracted by the vision of a glittering era of personalised genomic medicine, but fearful of the 'geneticized' future portrayed by the popular film GATTACA, in which people's employment prospects – and therefore much else about their place in society – would be directly determined by their genetic makeup, rather than their talents, passions and efforts.

It was very common in public meetings for participants to declare their basic altruism, and their willingness to provide tissue for specific genetic research studies or permanent biobanks – so long as they could be reassured about respect for their privacy, protection from subsequent discrimination (by insurers, employers and others) based on their genetic status, and the governance standards of the institution concerned (discussed further below). Thus, the ALRC saw its major challenge as the need to develop policies and practice which would both foster innovations in genetic research and practice that serve humanitarian ends, and provide sufficient comfort to the community that such work will always be subject to proper ethical scrutiny and legal (and other) controls.

Addressing Community Concerns

As mentioned, it was noted during the consultations that a strong spirit of altruism exists in the community, ready to be tapped in relation to human genetic research. However, policy-makers must address the associated, and very legitimate, concerns expressed by members of the community, or else risk a backlash that would dramatically set back all such research, however well-designed and intentioned.

A range of concerns were raised, but for these purposes may be grouped into four broad categories:

- ethical oversight;
- biobank governance;
- commercialization, access and equity, and benefit sharing; and
- protection from collateral damage: genetic discrimination.

Improved ethical oversight of genetic research

The 'Bible' on research ethics in Australia is the *National Statement on Ethical Conduct in Research Involving Humans* (the 'National Statement') (NHMRC 2007)[6] which was developed by the NHMRC, in association with the Australian Research Council (the ARC, the other major research funding body), and the Australian Vice-Chancellor's Committee (the AVCC, representing all Australian universities).

The National Statement contains ethical principles relevant to all research involving humans, but requires that particular matters are to be addressed where the research involves certain categories of vulnerable persons (such as Aboriginal and Torres Strait Islander people, children and persons with an intellectual or mental impairment) or may involve the use of radiation, assisted reproductive technology, clinical trials, epidemiology, human tissue samples or genetics. Critically, the National Statement sets out the formation, membership and functions of Human Research Ethics Committees (HRECs), which must review and approve all research projects involving human participants.

The primary function of an HREC is to protect the welfare and rights of participants in research. However, an often-overlooked secondary purpose of the National Statement, and thus of HRECs, is to 'facilitate research that is or will be of benefit to the researcher's community or to humankind' (NHMRC 2007: Preamble).

Responding to earlier criticisms about the perceived lack of independence, at least half the membership of an HREC must now be external to the institution, and sufficient incentives (especially in relation to public funding) have been built into the system to ensure a high degree of compliance. On the other hand, the ALRC

6 This document replaced the 1999 version of the National Statement, and was developed following extensive public consultation and debate. Although the National Statement is not legally binding, it stipulates that it must be used to inform the design, ethical review and conduct of human research that is funded by, or takes place under the auspices of, the NHMRC, the ARC or the AVCC. Compliance with the National Statement is a condition of NHMRC grants of research funding. In addition, in order for an institution to apply for or subsequently administer NHMRC research funds, all research conducted within the institution, whether or not funded by the NHMRC – such as clinical trials – must comply with the National Statement.

did identify a number of deficiencies, including that: many HRECs are overloaded and under-resourced; HREC members need better training and more support; there should be more transparency and accountability of processes; more meaningful monitoring needs to be done on an on-going basis; multi-centre research must be handled in a far more efficient manner; and, in some cases, HRECs still appear to lack sufficient independence from their host institutions.

Although AHEC has overall responsibility for monitoring the effectiveness of the system, there is no overarching approval framework. In common with the United Kingdom – but contrary to the position in Japan and most of Western Europe – Australia operates a decentralized, collegial system of over 220 HRECs, with no national level committee reviewing approvals or hearing appeals from refusals. However, the Australian approach has the benefits of being largely voluntary, flexible and relatively inexpensive, tapping local expertise and knowledge held within the institution – and low cost is a particular virtue in a system that does not usually allocate research funds to cover compliance with regulatory requirements.

In *Essentially Yours*, many recommendations were directed to the NHMRC and aimed at strengthening the mechanisms through which compliance with the National Statement is enforced (ALRC 2003: Chapters 14-17), including taking steps to better support the work of HRECs, such as providing induction programs for new members and training for continuing members. In order to promote greater transparency and consistency, and permit effective monitoring by AHEC, the ALRC recommended that the NHMRC develop a quality improvement framework for HRECs, and consider the introduction of a formal accreditation system for HRECs. The latest version of the National Statement (2007) addresses many of the concerns expressed in *Essentially Yours*, and follows the ALRC's suggested approach in providing dedicated chapters on databanks, human tissue samples and human genetics (NHMRC 2007: Chapters 3.2, 3.4 and 3.5).

Governance arrangements

Again, it was clear from the ALRC's public consultation effort that continuing community support is conditional upon its confidence in the probity and effectiveness of the governance of major genetic research projects and biobanks. Members of the community wanted assurance that the governance arrangements involve a high degree of independence, transparency, accountability, and financial and commercial integrity.

With regard to the latter, there is an appreciation that genetic research is an expensive business, requiring a commitment to public funding as well as a very significant level of private investment. Interestingly, it was not the profit motive or commercialization *per se* that worried most members of the community, but rather the fear of being treated less than honestly – there was considerable anger expressed by research participants who subsequently found out that their doctors or the research scientists involved had undisclosed financial interests in the project.

Any hint of a hidden monopoly or exclusive commercial arrangements – such as those revealed with relation to the first national biobank, established in Iceland – is likely to arouse great antipathy and to challenge the legitimacy of the entire project.

Empirical evidence about public attitudes to research in the United Kingdom is instructive. A survey conducted by the UK Human Genetics Commission found that levels of public trust in biobanking varied markedly, depending on the nature of the individuals or bodies holding the genetic information. In particular, respondents trusted academic scientists far more than pharmaceutical companies, and placed the lowest level of trust in 'government'.[7]

Similarly, another survey revealed that members of the public found it both reassuring and important in terms of credibility when a biobank has been established 'as a publicly funded initiative and not set up as a profit-making exercise' (Dawson 2003). The survey report also indicated that:

> [t]here are likely to be questions from the general public and in the media about commercial access to, and use of, the samples and information. Assuming samples are donated freely by donors, there needs to be careful explanation of the financial implications of this. (Dawson 2003: 17)

Leading Australian commentators also have expressed the view that, from an ethical perspective, 'the potential for commercial exploitation' of genetic samples and other biological materials is a very relevant consideration when individuals decide whether to consent to participate in research, given that participation is typically altruistic in nature (Nicol, Otlowski and Chalmers 2001).

The ALRC accepted that there is a clear need for open and transparent disclosure to prospective research participants of the potential commercialization of research outcomes and the commercial interests of the researchers involved. It was suggested that such disclosure may protect the interests of both prospective research participants and researchers themselves:

> In order to avoid feelings of exploitation, and possibly even deception, it is of crucial importance that they be given the opportunity to consent to participation in the knowledge that there is a possibility of commercial gain being made from their donated biological material. To do otherwise risks damaging the perception of research and may thereby reduce the willingness of people in the community to participate. (Nicol. Otlowski and Chalmers 2001: 93)

7 Respondents were asked the following question (n=1,038): Q68 Please tell me which, if any, you trust to use the human genetic information held on medical databases responsibly? The responses included: an expert government scientific advisory committee (39 per cent); academic scientists (38 per cent); health and pharmaceutical companies (20 per cent); government (13 per cent).

The Australian Academy of Sciences submitted to the ALRC's Genetic Information inquiry that 'all applications to HRECs should be required to include details of any commercial support obtained or envisaged' (ALRC 2003: Paragraph 16.47). Similarly, the Australian Red Cross Ethics Committee stated that:

> there should be transparent disclosure to research participants of the potential commercialisation of research outcomes, as well as any conflicts of interest. … [The] Committee requires disclosure of commercial arrangements for funding or product development. No researcher has ever raised any objection. In fact most researchers provide a full explanation of the commercial aspects of the research. We should also point out that our standard-model information form includes specific questions on commercialisation. There has been a general acceptance of the disclosure principle. (ALRC 2003: Paragraph 16.49)

The ALRC endorsed this approach and recommended that the NHMRC develop information and advice on disclosure by researchers to research participants of information about actual or anticipated commercial arrangements connected with human genetic research proposals (ALRC 2003: Recommendation 16-1) – and the National Statement now deals more fully with disclosure and the handling of conflicts of interest (NHMRC 2007: Chapter 5.4).

Access, equity and benefit sharing

There is also strong interest in the community in requiring the proponents of major research projects and biobanks to spell out the benefit sharing arrangements, and address issues of access and equity. Again, there is generally no expectation that volunteers will receive any direct financial remuneration or profit-sharing as a result of their participation in genetic research; however, there is a desire to be informed about the potential social or communal benefits.

Similarly, another commonly expressed fear at public meetings was that yet *another* major modern technology with the potential to make life better might instead drive up the costs of healthcare and increase the divide between the 'haves' and the 'have nots'. (This arose in the Privacy inquiry as well, with respect to the perceived 'digital divide' between city dwellers with relatively inexpensive broadband access, and those living in rural, regional and remote communities.) It was suggested that while 'smart drugs' based on modern pharmacogenomics should be a good thing, the resulting markets for individualised and customized drugs will be smaller and more fragmented – resulting in more effective drug therapies, but at much higher prices. Concerns were expressed, particularly by Aboriginal people, that this would tend to shape the research programs for drug companies, prompting them to focus more on 'white, middle class, lifestyle diseases' than on those diseases primarily affecting the poor, the disadvantaged and the marginalized.

Finally, people want to be presented with unbiased information about the relative benefits of the research project or biobank. Although there is generally strong support in the community for human genetic research, given the high cost and the potential risks involved, it is by no means universally accepted that it is wiser to focus on 'the gene for obesity' than on promoting good nutrition, or that biobanking is 'the most effective and cost-effective way to reduce the incidence of common diseases' (Wallace 2005: 34-36). As one commentator has noted, 'the only thing certain about these population-wide, genotype-phenotype resources is that they are staggeringly expensive' (Longtin 2004: 1567-1569, quoting Professor Hank Greely of Stanford Law School). In order to develop and maintain public support, those establishing biobanks must provide an honest assessment of the relative value and cost-effectiveness of this type of high-end research.

Protection from collateral damage: Genetic discrimination

The UNESCO *Universal Declaration on the Human Genome and Human Rights 1997* recognizes that:

> research on the human genome and the resulting applications open up vast prospects for progress in improving the health of individuals and of humankind as a whole, but ... that such research should fully respect human dignity, freedom and human rights, as well as the prohibition of all forms of discrimination based on genetic characteristics.

The fear of emerging genetic discrimination – especially in relation to insurance and employment – was central to the establishment of the ALRC's inquiry into Genetic Information. For example, researchers, doctors and community groups raised concerns that unless adequate legal protections were developed, people would become reluctant to volunteer for research studies or even to undergo needed clinical genetic tests, for fear of suffering adverse incidental consequences. For example, people might fear that a positive genetic test for a disease marker (such as Huntington's disease, or heritable breast or colorectal cancer) could be used to deny them risk-rated insurance coverage (or to elevate premiums beyond reach) or employment opportunities.

The law in Australia prohibits discrimination in a range of specified circumstances. At the federal level, the major laws include the *Sex Discrimination Act 1984* (Cth), the *Racial Discrimination Act 1975* (Cth), the *Disability Discrimination Act 1992* (Cth) (DDA) and the *Age Discrimination Act 2004* (Cth). In addition, the *Workplace Relations Act 1996* (Cth) prohibits discrimination on a range of grounds in respect of the termination of employment. However, at present, none of these Acts expressly address discrimination on the basis of genetic status. The DDA applies to unlawful discrimination based on a person's physical disability, mental illness, intellectual disability or HIV–AIDS positive status ('the presence in the body of organisms capable of causing disease or illness') – but was enacted before genetic discrimination emerged as a concern. In order to provide coverage and a consistent approach to addressing genetic discrimination, the ALRC recommended that the DDA and related laws and regulations[8] should be amended to make clear that they *expressly* apply to discrimination based on a person's real or perceived genetic status (ALRC 2003: Recommendation 9-3).

Further, the ALRC took a strongly interventionist approach to the use of genetic testing and information in the employment area, in order to forestall the creation of a 'genetic underclass' of people who are fit and willing to work, but who may show some predisposition to possible future illness, which employers might use to rule them out. The ALRC recommended that, as a general rule, employers should not be permitted to collect or use genetic information in relation to job applicants and employees. Exceptions to this should be permitted only in rare and compelling circumstances – for example, where such action is: (a) necessary to protect the health and safety of workers or third parties; *and* (b) complies with stringent standards developed for this purpose by the national advisory bodies on human genetics (the HGAC) and on occupational health and safety (ALRC 2003: Part H, especially Recommendation 30-1).

By way of contrast, in the area of risk-rated, private insurance (eg life insurance and critical illness insurance, but not private health insurance, which is 'community rated' in Australia), the ALRC inquiry did not favour intrusive government intervention at this time, preferring instead to call upon the industry and rely on the market to provide more effective regulation; greater adherence to scientific and medical advances (and their actuarial implications); more transparency, and greater responsiveness to legitimate consumer interests and concerns (ALRC 2003: Part G). However, the ALRC did recommend that the HGCA maintain 'a watching brief on developments in the insurance industry in relation to the use of human genetic information, both in Australia and overseas, with a view to reviewing Australian insurance practices as the need arises' (ALRC 2003: Recommendation 26-2).

The Australian Government accepted the great majority of these recommendations, including those in the employment area. Unfortunately, however, the legislative changes required to implement these policies have yet to be made. In the interim, similar concerns led to the recent passage of the *Genetic Information Non-Discrimination Act 2008* (GINA) in the United States by a unanimous vote in the Senate and 414-1 margin in the Congress.[9]

---

8 Such as the *Human Rights and Equal Opportunity Commission Act 1986* (Cth) and the *Workplace Relations Act 1996* (Cth). The ALRC also made a parallel recommendation that the states and territories should consider harmonizing their anti-discrimination legislation, and other relevant laws, in a manner consistent with the recommendations in the report: *Essentially Yours*, Recommendation 9-5.

9 The Bill was signed into law by President Bush on 21 May 2008.

Informed Consent, Privacy and Biobanks

'Informed consent'

The theory and practices surrounding 'informed consent' to medical procedures have long been vexed questions in bioethics, law and medicine. All researchers understand – and HRECs now require as a condition of approval – that special care must be taken to ensure that those providing tissue to a biobank, as with participants in any research project, have provided informed consent. As a matter of practice, this usually involves signing a consent form, following perusal of a document that sets out the nature of the research project, any medical risks involved, potential benefits and outcomes, and other material relevant for an individual making a full, free and informed decision to participate. The quality of the documentation has improved markedly in recent years – but it has also lengthened accordingly, raising a host of other questions about whether such consent truly can be meaningful in the circumstances.

When human genetic research reveals information important to the future health of an identified or potentially identifiable participant or his or her offspring, the research protocol must provide for the same consent, counselling and confidentiality protection as would apply in a clinical setting. In the particularly dynamic contexts of human genetic research and biobanking, it is now common practice for researchers to seek consent from donors for their tissue to be used in a particular experiment and then stored for possible later inclusion in other experiments, the details and potential implications of which are unspecified. This practice has been criticized in some quarters as inappropriate because, as a matter of principle, valid consent cannot be provided in the absence of full and complete disclosure of the specific uses to which the tissue is to be put.

However, there is also broad consensus within the medical research community about the need for procedures under which prospective research participants may provide consent for *unspecified* future research, provided that such research is carefully scrutinized and approved by an HREC – which may, in some circumstances, require researchers to go back and obtain fresh consent from the donors. Suggestions for appropriate procedures often focus on the idea of 'tick a box' consent options; that is, prospective research participants could select from a graduated set of consent options, so that in addition to participants being asked to consent to participation in the specified research study, they also could opt to have their tissue or information stored for 'related' or 'unrelated' future research, and designate whether future use would, or would not, require fresh consent. Similarly, it is sometimes suggested that the language of consent should be augmented by the concept of 'donation' or 'gifting'; that is, altruistic individuals may wish to 'gift' samples for research purposes without condition – for example by 'leaving their body to science' or donating blood to the Red Cross.

These aims are now substantially achieved in the 2007 edition of the National Statement, in section 3.2 on Databanks, with section 3.2.9 providing guidance on consent issues in this specific context, in association with sections 2.2 and 2.3, which set out the general requirements for consent and for qualifying or waiving conditions for consent (respectively). However, more work is still needed and creative thought must be given to adapting the traditional approaches to 'informed consent' to the new circumstances and exigencies of the genomics era. Mascalzoni and others have argued persuasively that informed consent can no longer remain 'a one-off affair', with the nature of the genomic research now requiring the addition of a temporal dimension: as long as researchers have access to an individual's DNA, they should be obliged to maintain an ongoing relationship and interactions. Rather than a single broad consent, they propose a 'circular process of information exchange' involving public presentations, feedback from research subjects, and information being provided by researchers to subjects (if desired) – which can be facilitated through the clever use of information and communication technology, such as email, mobile phone messaging, regularly updated Web pages and social networking websites (Mascalzoni et al. 2008. Discussed in Anderson, A. 2008. *Genomics Needs Informed Consent Overhaul, European Researchers Say*. [Online] Available at: http://www.genomeweb.com/issues/news/149408-1.html).

The ALRC recommended that the NHMRC develop the National Statement to provide express ethical guidance on biobanking – and, given the particular nature and patterns of genetic research, to offer ethical and practical guidance on obtaining consent to unspecified future research. The ALRC also recommended that the National Statement provide guidance on the use and relative efficacy of such strategies as anonymization, de-identification and encryption to protect the privacy and confidentiality of volunteers; and 'gene trustee' or 'genetic ombudsman' mechanisms, designed to insure there is an independent element in the process to increase privacy protection and oversee the integrity of consent agreements.

Research and waivers of consent under privacy law

In Australia, the *Privacy Act* regime incorporates the HREC approval process established by the National Statement to ensure that wherever research is conducted using personal information without consent, that research pays due regard for the balance of public interests and the protection of personal information. In common with the privacy laws of other comparable countries, the Australian *Privacy Act* recognizes that personal 'health information' should be designated as 'sensitive information', in most cases deserving the highest level of protection (*Privacy Act 1988*: s.6(1)). In accordance with recommendations made in *Essentially Yours*, the *Privacy Act* was amended in 2006 to make clear that 'health information' includes genetic information.[10]

---

10 The *Privacy Legislation Amendment Act 2006* (Cth) inserted into s.6(1) a definition of 'genetic relative'; amended the definition of 'health information' to cover genetic information predictive of the health of the individual or a genetic relative (now s.6(1)(d));

In a submission to the ALRC's Privacy inquiry, the Commonwealth Scientific and Industrial Research Organisation (CSIRO) noted that:

> Informed consent and opt-in is a good model for clinical trials, for example, where the risk is normally predominantly to the participating individual. However, in the case of population health research, the findings will often be implemented for the whole population. In these cases informed consent and opt-in may not be good models because non-participation can introduce bias and therefore affect the applicability of the results. (ALRC 2008: Paragraph 64.24)

Other submissions and research provided to the ALRC's Privacy inquiry confirmed the impressions of community attitudes gained during the earlier work on Genetic Information. For example, research by the federal Department of Health and Ageing (DoHA) suggested that, while patients were concerned about the disclosure of identified personal information for purposes other than their own clinical care, they generally are very accepting of de-identified health information being shared amongst health planners and researchers (DOHA 2004. See ALRC 2008: Paragraph 64.21). Similarly, research conducted by the NHMRC indicated that there was considerable support among the general public (66 per cent) and health consumers (64 per cent) for approved researchers to match information from different databases, and there was an even higher level of support for approved researchers to access health information from databases where this was identified by a unique number rather than a name (NHMRC 2004. See ALRC 2008: Paragraph 64.21). In to a survey of patients attending an oncology clinic at the Royal Adelaide Hospital in 2005, 93 per cent reported that they would allow their health information to be used for research, so long as it was kept confidential and they could not be identified. Where the health information was identifiable, 33 per cent would nevertheless allow the information to be used without their express consent. A further 15 per cent would allow the use of the information without consent if the research project was HREC-approved (as it must be in any event), and 6 per cent would allow the use of identifiable information without their consent if it was impracticable to obtain such consent (Beeke, Olver and McLaughlin 2007). In a joint submission to the ALRC following up these survey results, Cancer Council Australia and the Clinical Oncological Society of Australia suggested that there would be even more support for the use of health information in these circumstances if there was a greater appreciation in the community of how such research contributes to improvements in cancer prevention, detection and treatment – and also that the level of support would increase if the community was better informed about the mechanisms available to protect privacy (ALRC 2008: Paragraph 64.23).

---

and amended the definition of 'sensitive information' to include 'genetic information about an individual that is not otherwise health information' (now s.6(1)(c)).

Indeed, both the National Statement and the *Privacy Act* acknowledge that in some circumstances it is very difficult or impossible to conduct research in a way that complies with the various Privacy Principles enshrined in the Act, but that nevertheless may be in the public interest – for example, epidemiological studies of the distribution and determinants of disease in large populations. Consequently, the *Privacy Act* provides a mechanism to allow health and medical research to go forward without consent, subject to Guidelines issued by the NHMRC under sections 95 (in relation to public sector agencies) (NHMRC 2000) and 95A (in relation to organizations) (NHMRC 2001)[11] and approved by the Privacy Commissioner.

Basically – and contrary to a good deal of mythology about the deadening effect of privacy law on research – the existing law[12] provides that health information may be collected, used and disclosed where necessary for research or the compilation or analysis of statistics, relevant to public health or public safety where:

- the purpose cannot be served by the collection of information that does not identify the individual;
- it is impracticable for the organization to seek the individual's consent to the collection, use or disclosure;
- the information is collected, used and disclosed in accordance with rules approved by the Privacy Commissioner;
- in the case of disclosure – the organization reasonably believes that the recipient of the health information will not disclose the health information, or personal information derived from the health information; and
- the organization takes reasonable steps to permanently de-identify the information before it discloses it.

These Guidelines provide HRECs with a detailed framework for considering the privacy implications of health and medical research proposals involving the use of individuals' personal or health information. Importantly, HRECs are authorized to approve research proposals seeking to use such identifiable information *without consent* where the public interest in the research 'substantially outweighs' the public interest in maintaining the level of privacy protection provided by the *Privacy Act*. In striking this balance, HRECs are asked to have regard to a long list of factors, including:

- the degree to which the personal information is necessary for the research;
- the public importance of the research and the likely contribution to the community;

---

11 One of the central concerns in *For Your Information* (ALRC 108) is to streamline and harmonize Australian privacy laws, including consolidation of the two separate sets of Privacy Principles into a single set that applies to both the public and private sectors.

12 National Privacy Principles (NPP) 2 and 10, operating in combination.

- any likely benefits to individuals or groups;
- whether the research could be achieved within the terms of the Privacy Principles and the degree to which this would impact on the scientific value of the research;
- whether the risk of harm to the individual whose personal information is to be used is minimal;
- the study design and scientific credentials of those involved in the research;
- whether access to the information is restricted to appropriate personnel;
- the procedures to be followed to ensure that the information is permanently de-identified before the publication of results; and
- the procedures to be followed at the completion of the study to protect or destroy the information.

Although the evidence put before the ALRC was not strong, it suggested that the current test – especially insofar as it requires an HREC to find that the public interest in the research 'substantially outweighs' the interest in privacy protection – may lead to unduly cautious decision-making. The NHMRC submitted that the 'substantially outweighs' requirement:

> is unbalanced and is limiting the conduct of important health and medical research … In undertaking an assessment for the purposes of determining the balance of public interests, an HREC routinely assesses a range of issues, which are detailed [above]. This assessment provides a robust framework and in our view protects the reasonable interests of individuals. It is clear that an assessment would not favour research that has the potential to cause significant harm to individuals. We consider that a more appropriate and effective test that would accord with community sentiment would simply be that the balance of public interests favours the research proceeding. (ALRC 2008: Paragraph 65.74)

Whatever the current empirical reality with respect to HREC decision-making, the ALRC preferred the NHMRC's approach in policy terms, and recommended that the test be modified, so that in future an HREC only must be satisfied that 'the public interest in the activity outweighs the public interest in maintaining the level of privacy protection provided by the *Privacy Act*' (ALRC 2008: Recommendation 65-4).

The ALRC also considered submissions that suggested there is some uncertainty surrounding the current waiver requirement that it must be 'impracticable for the organization to seek the individual's consent'. For example, one HREC argued that researchers and HRECs find the term confusing, and wondered whether guidance was needed on when the cost of obtaining consent would make it 'impracticable' (ALRC 2008: Paragraph 65.92). The NHMRC submitted that:

> It is not clear to us whether it would be considered impracticable to seek consent in circumstances where research subjects are contactable but the process of seeking consent would damage the scientific integrity of the proposed research. In addition, if a research participant previously has given consent in general terms for their health information to be used in a future similar research study, even though it may not be 'impracticable' to seek specific consent for the second study it may be quite unnecessary and inefficient to do so. (ALRC 2008: Paragraph 65.92)

The ALRC agreed and recommended accordingly (ALRC 2008: Recommendation 65-5). Together with a number of other small changes to content and wording, the net result of the ALRC's recommended 'Research Exception' to the Privacy Principles reads as follows:

> *Recommendation 65-8.* The research exception to the 'Collection' principle should provide that an agency or organisation may collect personal information, including sensitive information, about an individual where all of the following conditions are met:
>
> (a) the collection is necessary for research;
>
> (b) the purpose cannot be served by the collection of information that does not identify the individual;
>
> (c) it is unreasonable or impracticable for the agency or organisation to seek the individual's consent to the collection;
>
> (d) a Human Research Ethics Committee – constituted in accordance with, and acting in compliance with, the *National Statement on Ethical Conduct in Human Research* as in force from time to time – has reviewed the proposed activity and is satisfied that the public interest in the activity outweighs the public interest in maintaining the level of privacy protection provided by the *Privacy Act*; and
>
> (e) the information is collected in accordance with the Research Rules, to be issued by the Privacy Commissioner.
>
> Where an agency or organisation collects personal information about an individual under this exception, it must take reasonable steps to ensure that the information is not disclosed in a form that would identify the individual or from which the individual would be reasonably identifiable.
>
> *Recommendation 65-9.* The research exception to the 'Use and Disclosure' principle should provide that an agency or organisation may use or disclose personal information where all of the following conditions are met:

(a) the use or disclosure is necessary for research;

(b) it is unreasonable or impracticable for the agency or organisation to seek the individual's consent to the use or disclosure;

(c) a Human Research Ethics Committee – constituted in accordance with, and acting in compliance with, the *National Statement on Ethical Conduct in Human Research* as in force from time to time – has reviewed the proposed activity and is satisfied that the public interest in the activity outweighs the public interest in maintaining the level of privacy protection provided by the *Privacy Act*;

(d) the information is used or disclosed in accordance with the Research Rules, to be issued by the Privacy Commissioner; and

(e) in the case of disclosure – the agency or organisation reasonably believes that the recipient of the personal information will not disclose the information in a form that would identify the individual or from which the individual would be reasonably identifiable.

Conclusions

Genetic researchers have already picked much of the available 'low-lying fruit', identifying monogenic conditions and others involving relatively simple associations. Further advances in scientific and medical knowledge – especially in relation to understanding polygenic disorders and the complex relationship between genes and environment – increasingly will involve population genetics, with researchers requiring access to large collections of stored genetic material and to genetic registers and databases. Rapid advances in genetic science and technology in recent years increasingly makes such analysis possible. PCR machines permit the amplification of small amounts of genetic material, so that it is possible to produce an almost inexhaustible supply of testable material once a sample has been obtained. Microarrays permit researchers to inspect and record many thousands of regions of the genome simultaneously. Supercomputers crunch the vast amounts of data so obtained, reducing to a matter of seconds the sort of painstaking analytical work that previously may have taken years. These advances also greatly challenge the traditional paradigm of individual informed consent, in which a person is asked to consider the information provided and decide whether or not to participate in a defined and limited research project or clinical trial. As Wjst and others have also noted, the nature of the risk to participants in population genetics research is also qualitatively different – whereas volunteers in a clinical trial run a risk of physical harm or suffering, participants in genomic research may be exposed to a range of collateral 'new threats to the individual such as uninsurability, unemployability, genetic discrimination, or disruption of family relationships' (Mascalzoni et al. 2008. Discussed in Anderson, A. 2008. *Genomics Needs Informed Consent Overhaul, European Researchers Say*. [Online] Available at: http://www.genomeweb.com/issues/news/149408-1.html).

How does the ethical researcher now design a protocol and consent form to explain to the ordinary member of the community the myriad contingencies relating to the nature of the research, and its risks and benefits, as described above? The framework for ethical oversight and research management in Australia has begun evolving to adapt to the nature and needs of human genetic research. The revised National Statement now provides specific guidance on such research and on biobanking. Although many researchers don't understand the *Privacy Act* – viewing it as a blunt instrument that hinders creative research – it actually already allows for common sense and flexibility in this area, and if the ALRC's recent recommendations (as advocated by the NHMRC and others) are implemented, this would improve the situation further.

Public surveys and community consultation in Australia and other countries generally indicate a high degree of support for genetic research and medicine, with an appropriate level of awareness about the need for effective ethical oversight and legal regulation of such research, as well as concerns about potential social harms, including increased discrimination and loss of personal privacy and inequities in access to the healthcare. It is now essential to promote an open, honest, intelligent and wide-ranging discussion about the ethics we need in this new era of population genetics and biobanking, which may require us to rearticulate the balance between individual privacy and consent on the one hand, and the broader public health benefits that may be gained from research on our 99.9 per cent shared human genome. A precondition to asking individuals to surrender any degree of personal autonomy and control for the common good must be the development of laws and policies that provide assurance that their concerns – about privacy, security, discrimination, governance, commercial fairness, financial probity, and access and equity in relation to genetic medicine – will be addressed in structural terms and monitored thereafter. Given the natural suspicion of those with vested interests in promoting genetic research, we will need independent, 'trusted voices' – as provided by respected bodies like the UK Human Genetics Commission – to help frame and lead these debates.

The key lesson emerging from biobanking and population genetics initiatives to date is that there is a clear need for openness and transparency in order to maintain public confidence. This experience amply demonstrates that taking ethics and governance seriously – and engaging the community in the discussion from an early stage, and thereafter – is essential to achieving acceptance and legitimacy.

Bibliography

ABC News. 2000 *Clean Genes: Genetically Pure Polynesian Paradise Pegged for Research*. [Online] Available at: http://abcnews.go.com/sections/living/DailyNews/genes_tonga001122.html [accessed 22 November 2000].

Alboim, E. 2005. Public Attitudes Towards Genomics in Europe and North America, in *Genomics, Health and Society: Emerging Issues for Public Policy*, Health Canada (Symposium Report, 24-25 March 2005), 23-7.

Australian Government Department of Health and Ageing, *Submission to the Office of the Privacy Commissioner Review of the Private Sector Provisions of the Privacy Act 1988*, December 2004.

Australian Law Reform Commission. 2003. *Essentially Yours: The Protection of Human Genetic Information in Australia* (ALRC 96).

Australian Law Reform Commission. 2008. *For Your Information: Privacy Law and Practice* (ALRC 108).

Beeke, C., Olver, I. and McLaughlin, K. 2007.A Survey of Patients' Attitudes Towards the Use of their Health Data. *Journal of Registry Management*, 34(4), 119.

Burton, B. 2002. Proposed Genetic Database on Tongans Opposed. *British Medical Journal*, 324(7335), 443.

Dawson, C.R. 2003. *Public Perceptions of the Collection of Human Biological Samples*. [Online] Available at: MRC and Wellcome Trust, www.mrc.ac.uk/pdf-biobank_public_perceptions.pdf [accessed 20 February 2003].

Eurobarometer 52.1. 2003. *The Europeans and Biotechnology*. [Online] Available at: www.europa.eu.int/comm/research/quality-of-life/eurobarometer.html.

Howe, M. 2001. Australian Company Buys Rights to Tonga Gene Pool. *The Lancet Oncology*, 2(1).

Longtin, R. 2004. Canadian Province Seeks Control of Its Genes. *Journal of the National Cancer Institute*, 96, 1567-69.

Mascalzoni, D., Hicks, A., Pramstaller, A. and Wjst, M. 2008. Informed Consent in the Genomics Era. *PloS Medicine*, 5(9), 192.

Mori Social Research. 2003. *Public Attitudes to Human Genetic Information: People's Panel Quantitative Study Conducted for the Human Genetics Commission*. [Online] Available at: www.hgc.gov.uk/business_publications_morigeneticattitudes.pdf.National, Health and Medical Research Council, *Guidelines under Section 95 of the Privacy Act 1988* (2000).

National Health and Medical Research Council. 2001. *Guidelines Approved under Section 95A of the Privacy Act 1988*. Canberra: NHMRC.

National Health and Medical Research Council. 2004. *Submission to the Office of the Privacy Commissioner Review of the Private Sector Provisions of the Privacy Act 1988*. Canberra: NHMRC.

National Health and Medical Research Council. 2007. *National Statement on Ethical Conduct in Research Involving Humans*. Canberra: NHMRC.

Nicol, D., Otlowski, M. and Chalmers, D. 2001. Consent, Commercialisation and Benefit-Sharing. *Journal of Law and Medicine*, 9(1), 80.

Privacy Act 1988 (Cth).

Skatssoon, J. 2006. *Genetic Databank to Catalogue WA Residents*. [Online] Available at: ABC Science Online, http://www.abc.net.au/news/newsitems/200608/s1709816.htm.

UNESCO. *Universal Declaration on the Human Genome and Human Rights*. [Online] Available at: www.unesco.org/ibc/en/genome/projet.

Wallace, H. 2005. *Genomics, Health and Society: Emerging Issues for Public Policy*, Health Canada (Symposium Report, 24-25 March 2005), 34-6.

Index